Java Web Services: Up and Running

Java Web Services: Up and Running

Martin Kalin

O'REILLY®

Beijing · Cambridge · Farnham · Köln · Sebastopol · Tokyo

Java Web Services: Up and Running

by Martin Kalin

Published by O'Reilly Media, Inc., 1005 Gravenstein Highway North, Sebastopol, CA 95472.

O'Reilly books may be purchased for educational, business, or sales promotional use. Online editions are also available for most titles (*http://safari.oreilly.com*). For more information, contact our corporate/institutional sales department: 800-998-9938 or *corporate@oreilly.com*.

Editors: Mike Loukides and Julie Steele
Production Editor: Sarah Schneider
Production Services: Appingo, Inc.

Cover Designer: Karen Montgomery
Interior Designer: David Futato
Illustrator: Robert Romano

Printing History:
    February 2009:   First Edition.

ISBN: 978-0-596-52112-7

[SB]   [11/10]

1287353000

Table of Contents

Preface

This is a book for programmers interested in developing Java web services and Java clients against web services, whatever the implementation language. The book is a code-driven introduction to *JAX-WS* (Java API for XML-Web Services), the framework of choice for Java web services, whether SOAP-based or REST-style. My approach is to interpret JAX-WS broadly and, therefore, to include leading-edge developments such as the Jersey project for REST-style web services, officially known as *JAX-RS* (Java API for XML-RESTful Web Services).

JAX-WS is bundled into the *Metro Web Services Stack*, or *Metro* for short. Metro is part of core Java, starting with Standard Edition 6 (hereafter, core Java 6). However, the Metro releases outpace the core Java releases. The current Metro release can be downloaded separately from *https://wsit.dev.java.net*. Metro is also integrated into the Sun application server, GlassFish. Given these options, this book's examples are deployed in four different ways:

Core Java only
> This is the low-fuss approach that makes it easy to get web services and their clients up and running. The only required software is the Java software development kit (SDK), core Java 6 or later. Web services can be deployed easily using the `Endpoint`, `HttpServer`, and `HttpsServer` classes. The early examples take this approach.

Core Java with the current Metro release
> This approach takes advantage of Metro features not yet available in the core Java bundle. In general, each Metro release makes it easier to write web services and clients. The current Metro release also indicates where JAX-WS is moving. The Metro release also can be used with core Java 5 if core Java 6 is not an option.

Standalone Tomcat
> This approach builds on the familiarity among Java programmers with standalone web containers such as Apache Tomcat, which is the reference implementation. Web services can be deployed using a web container in essentially the same way as are servlets, JavaServer Pages (JSP) scripts, and JavaServer Faces (JSF) scripts. A standalone web container such as Tomcat is also a good way to introduce container-managed security for web services.

GlassFish

This approach allows deployed web services to interact naturally with other enterprise components such as Java Message Service topics and queues, a *JNDI* (Java Naming and Directory Interface) provider, a backend database system and the `@Entity` instances that mediate between an application and the database system, and an *EJB* (Enterprise Java Bean) container. The EJB container is important because a web service can be deployed as a stateless Session EJB, which brings advantages such as container-managed thread safety. GlassFish works seamlessly with Metro, including its advanced features, and with popular *IDE*s (Integrated Development Environment) such as NetBeans and Eclipse.

An appealing feature of JAX-WS is that the API can be separated cleanly from deployment options. One and the same web service can be deployed in different ways to suit different needs. Core Java alone is good for learning, development, and even lightweight deployment. A standalone web container such as Tomcat provides additional support. A Java application server such as GlassFish promotes easy integration of web services with other enterprise technologies.

Code-Driven Approach

My code examples are short enough to highlight key features of JAX-WS but also realistic enough to show off the production-level capabilities that come with the JAX-WS framework. Each code example is given in full, including all of the `import` statements. My approach is to begin with a relatively sparse example and then to add and modify features. The code samples vary in length from a few statements to several pages of source. The code is deliberately modular. Whenever there is a choice between conciseness and clarity in coding, I try to opt for clarity.

The examples come with instructions for compiling and deploying the web services and for testing the service against sample clients. This approach presents the choices that JAX-WS makes available to the programmer but also encourages a clear and thorough analysis of the JAX-WS libraries and utilities. My goal is to furnish code samples that can serve as templates for commercial applications.

JAX-WS is a rich API that is explored best in a mix of overview and examples. My aim is to explain key features about the architecture of web services but, above all, to illustrate each major feature with code examples that perform as advertised. Architecture without code is empty; code without architecture is blind. My approach is to integrate the two throughout the book.

Web services are a modern, lightweight approach to *distributed software systems*, that is, systems such as email or the World Wide Web that require different software components to execute on physically distinct devices. The devices can range from large servers through personal desktop machines to handhelds of various types. Distributed systems are complicated because they are made up of networked components. There is nothing more frustrating than a distributed systems example that does not work as claimed because the debugging is tedious. My approach is thus to provide full, working examples together with short but precise instructions for getting the sample application up and running. All of the source code for examples is available from the book's companion site, at *http://www.oreilly.com/catalog/9780596521127*. My email address is *kalin@cdm.depaul.edu*. Please let me know if you find any code errors.

Chapter-by-Chapter Overview

The book has seven chapters, the last of which is quite short. Here is a preview of each chapter:

Chapter 1, *Java Web Services Quickstart*
> This chapter begins with a working definition of *web services*, including the distinction between SOAP-based and REST-style services. This chapter then focuses on the basics of writing, deploying, and consuming SOAP-based services in core Java. There are web service clients written in Perl, Ruby, and Java to underscore the language neutrality of web services. This chapter also introduces Java's SOAP API and covers various ways to inspect web service traffic at the wire level. The chapter elaborates on the relationship between core Java and Metro.

Chapter 2, *All About WSDLs*
> This chapter focuses on the service contract, which is a *WSDL* (Web Service Definition Language) document in SOAP-based services. This chapter covers the standard issues of web service style (`document` versus `rpc`) and encoding (`literal` versus `encoded`). This chapter also focuses on the popular but unofficial distinction between the *wrapped* and *unwrapped* variations of document style. All of these issues are clarified through examples, including Java clients against Amazon's E-Commerce services. This chapter explains how the *wsimport* utility can ease the task of writing Java clients against commercial web services and how the *wsgen* utility figures in the distinction between `document`-style and `rpc`-style web services. The basics of *JAX-B* (Java API for XML-Binding) are also covered. This chapter, like the others, is rich in code examples.

Chapter 3, *SOAP Handling*
> This chapter introduces SOAP and logical handlers, which give the service-side and client-side programmer direct access to either the entire SOAP message or just its payload. The structure of a SOAP message and the distinction between SOAP 1.1 and SOAP 1.2 are covered. The messaging architecture of a SOAP-based service is discussed. Various code examples illustrate how SOAP messages can be processed in support of application logic. This chapter also explains how transport-level messages (for instance, the typical HTTP messages that carry SOAP payloads in SOAP-based web services) can be accessed and manipulated in JAX-WS. This chapter concludes with a section on JAX-WS support for

transporting binary data, with emphasis on *MTOM* (Message Transmission Optimization Mechanism).

Chapter 4, *RESTful Web Services*

This chapter opens with a technical analysis of what constitutes a REST-style service and moves quickly to code examples. The chapter surveys various approaches to delivering a Java-based RESTful service: `WebServiceProvider`, `HttpServlet`, Jersey Plain Old Java Object (POJO), and restlet among them. The use of a *WADL* (Web Application Definition Language) document as a service contract is explored through code examples. The *JAX-P* (Java API for XML-Processing) packages, which facilitate XML processing, are also covered. This chapter offers several examples of Java clients against real-world REST-style services, including services hosted by Yahoo!, Amazon, and Tumblr.

Chapter 5, *Web Services Security*

This chapter begins with an overview of security requirements for real-world web services, SOAP-based and REST-style. The overview covers central topics such as mutual challenge and message confidentiality, users-roles security, and WS-Security. Code examples clarify transport-level security, particularly under HTTPS. Container-managed security is introduced with examples deployed in the standalone Tomcat web container. The security material introduced in this chapter is expanded in the next chapter.

Chapter 6, *JAX-WS in Java Application Servers*

This chapter starts with a survey of what comes with a Java Application Server (JAS): an EJB container, a messaging system, a naming service, an integrated database system, and so on. This chapter has a variety of code examples: a SOAP-based service implemented as a stateless Session EJB, `WebService` and `WebServiceProvider` instances deployed through embedded Tomcat, a web service deployed together with a traditional website application, a web service integrated with *JMS* (Java Message Service), a web service that uses an `@Entity` to read and write from the Java DB database system included in GlassFish, and a WS-Security application under GlassFish.

Chapter 7, *Beyond the Flame Wars*

This is a very short chapter that looks at the controversy surrounding SOAP-based and REST-style web services. My aim is to endorse both approaches, either of which is superior to what came before. This chapter traces modern web services from DCE/RPC in the early 1990s through CORBA and DCOM up to the Java EE and .NET frameworks. This chapter explains why either approach to web services is better than the distributed-object architecture that once dominated in distributed software systems.

Freedom of Choice: The Tools/IDE Issue

Java programmers have a wide choice of productivity tools such as Ant and Maven for scripting and IDEs such as Eclipse, NetBeans, and IntelliJ IDEA. Scripting tools and IDEs increase productivity by hiding grimy details. In a production environment, such tools and IDEs are the sensible way to go. In a learning environment, however, the goal is to understand the grimy details so that this understanding can be brought to good use during the inevitable bouts of debugging and application maintenance. Accordingly, my book is neutral with respect to scripting tools and IDEs. Please feel free to use whichever tools and IDE suit your needs. My how-to segments go over code compilation, deployment, and execution at the command line so that details such as classpath inclusions and compilation/execution flags are clear. Nothing in any example depends on a particular scripting tool or IDE.

Conventions Used in This Book

The following typographical conventions are used in this book:

Italic
> Indicates new terms, URLs, filenames, file extensions, and emphasis.

`Constant width`
> Used for program listings as well as within paragraphs to refer to program elements such as variable or method names, data types, environment variables, statements, and keywords.

`Constant width bold`
> Used within program listings to highlight particularly interesting sections and in paragraphs to clarify acronyms.

 This icon signifies a tip, suggestion, or general note.

Using Code Examples

This book is here to help you get your job done. In general, you may use the code in this book in your programs and documentation. You do not need to contact us for permission unless you're reproducing a significant portion of the code. For example, writing a program that uses several chunks of code from this book does not require permission. Selling or distributing a CD-ROM of examples from O'Reilly books does require permission. Answering a question by citing this book and quoting example code does not require permission. Incorporating a significant amount of example code from this book into your product's documentation does require permission.

We appreciate, but do not require, attribution. An attribution usually includes the title, author, publisher, and ISBN. For example: *"Java Web Services: Up and Running*, by Martin Kalin. Copyright 2009 Martin Kalin, 978-0-596-52112-7."

If you feel your use of code examples falls outside fair use or the permission given above, feel free to contact us at *permissions@oreilly.com*.

Safari® Books Online

When you see a Safari® Books Online icon on the cover of your favorite technology book, that means the book is available online through the O'Reilly Network Safari Bookshelf.

Safari offers a solution that's better than e-books. It's a virtual library that lets you easily search thousands of top tech books, cut and paste code samples, download chapters, and find quick answers when you need the most accurate, current information. Try it for free at *http://safari.oreilly.com*.

How to Contact Us

Please address comments and questions concerning this book to the publisher:

O'Reilly Media, Inc.
1005 Gravenstein Highway North
Sebastopol, CA 95472
800-998-9938 (in the United States or Canada)
707-829-0515 (international or local)
707-829-0104 (fax)

We have a web page for this book, where we list errata, examples, and any additional information. You can access this page at:

http://www.oreilly.com/catalog/9780596521127/

To comment or ask technical questions about this book, send email to:

bookquestions@oreilly.com

For more information about our books, conferences, Resource Centers, and the O'Reilly Network, see our website at:

http://www.oreilly.com/

Acknowledgments

Christian A. Kenyeres, Greg Ostravich, Igor Polevoy, and Ken Yu were kind enough to review this book and to offer insightful suggestions for its improvement. They made the book better than it otherwise would have been. I thank them heartily for the time and effort that they invested in this project. The remaining shortcomings are mine alone, of course.

I'd also like to thank Mike Loukides, my first contact at O'Reilly Media, for his role in shepherding my initial proposal through the process that led to its acceptance. Julie Steele, my editor, has provided invaluable support and the book would not be without her help. My thanks go as well to the many behind-the-scenes people at O'Reilly Media who worked on this project.

This book is dedicated to Janet.

## CHAPTER 1

Java Web Services Quickstart

What Are Web Services?

Although the term *web service* has various, imprecise, and evolving meanings, a glance at some features typical of web services will be enough to get us into coding a web service and a client, also known as a consumer or requester. As the name suggests, a web service is a kind of *webified application*, that is, an application typically delivered over *HTTP* (Hyper Text Transport Protocol). A web service is thus a distributed application whose components can be deployed and executed on distinct devices. For instance, a stock-picking web service might consist of several code components, each hosted on a separate business-grade server, and the web service might be consumed on PCs, handhelds, and other devices.

Web services can be divided roughly into two groups, *SOAP*-based and *REST*-style. The distinction is not sharp because, as a code example later illustrates, a SOAP-based service delivered over HTTP is a special case of a REST-style service. *SOAP* originally stood for Simple Object Access Protocol but, by serendipity, now may stand for Service Oriented Architecture (SOA) Protocol. Deconstructing SOA is nontrivial but one point is indisputable: whatever SOA may be, web services play a central role in the SOA approach to software design and development. (This is written with tongue only partly in cheek. SOAP is officially no longer an acronym, and SOAP and SOA can live apart from one another.) For now, SOAP is just an *XML* (EXtensible Markup Language) dialect in which documents are messages. In SOAP-based web services, the SOAP is mostly unseen infrastructure. For example, in a typical scenario, called the request/response *message exchange pattern* (MEP), the client's underlying SOAP library sends a SOAP message as a service request, and the web service's underlying SOAP library sends another SOAP message as the corresponding service response. The client and the web service source code may provide few hints, if any, about the underlying SOAP (see Figure 1-1).

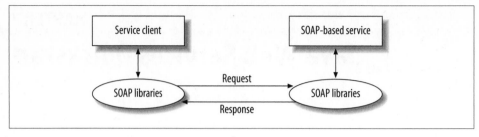

Figure 1-1. Architecture of a typical SOAP-based web service

REST stands for REpresentational State Transfer. Roy Fielding, one of the main authors of the HTTP specification, coined the acronym in his Ph.D. dissertation to describe an architectural style in the design of web services. SOAP has standards (under the World Wide Web Consortium [W3C]), toolkits, and bountiful software libraries. REST has no standards, few toolkits, and meager software libraries. The REST style is often seen as an antidote to the creeping complexity of SOAP-based web services. This book covers SOAP-based and REST-style web services, starting with the SOAP-based ones.

Except in test mode, the client of either a SOAP-based or REST-style service is rarely a web browser but rather an application without a graphical user interface. The client may be written in any language with the appropriate support libraries. Indeed, a major appeal of web services is language transparency: the service and its clients need not be written in the same language. Language transparency is the key to web service *interoperability*; that is, the ability of web services and requesters to interact seamlessly despite differences in programming languages, support libraries, and platforms. To underscore this appeal, clients against our Java web services will be written in various languages such as C#, Perl, and Ruby, and Java clients will consume services written in other languages, including languages unknown.

There is no magic in language transparency, of course. If a SOAP-based web service written in Java can have a Perl or a Ruby consumer, there must be an intermediary that handles the differences in data types between the service and the requester languages. XML technologies, which support structured document interchange and processing, act as the intermediary. For example, in a typical SOAP-based web service, a client transparently sends a SOAP document as a request to a web service, which transparently returns another SOAP document as a response. In a REST-style service, a client might send a standard HTTP request to a web service and receive an appropriate XML document as a response.

Several features distinguish web services from other distributed software systems. Here are three:

Open infrastructure

> Web services are deployed using industry-standard, vendor-independent protocols such as HTTP and XML, which are ubiquitous and well understood. Web services

can piggyback on networking, data formatting, security, and other infrastructures already in place, which lowers entry costs and promotes interoperability among services.

Language transparency
Web services and their clients can interoperate even if written in different programming languages. Languages such as C/C++, C#, Java, Perl, Python, Ruby, and others provide libraries, utilities, and even frameworks in support of web services.

Modular design
Web services are meant to be modular in design so that new services can be generated through the integration and layering of existing services. Imagine, for example, an inventory-tracking service integrated with an online ordering service to yield a service that automatically orders the appropriate products in response to inventory levels.

What Good Are Web Services?

This obvious question has no simple, single answer. Nonetheless, the chief benefits and promises of web services are clear. Modern software systems are written in a variety of languages—a variety that seems likely to increase. These software systems will continue to be hosted on a variety of platforms. Institutions large and small have significant investment in legacy software systems whose functionality is useful and perhaps mission critical; and few of these institutions have the will and the resources, human or financial, to rewrite their legacy systems.

It is rare that a software system gets to run in splendid isolation. The typical software system must interoperate with others, which may reside on different hosts and be written in different languages. Interoperability is not just a long-term challenge but also a current requirement of production software.

Web services address these issues directly because such services are, first and foremost, language- and platform-neutral. If a legacy COBOL system is exposed through a web service, the system is thereby interoperable with service clients written in other programming languages.

Web services are inherently *distributed* systems that communicate mostly over HTTP but can communicate over other popular transports as well. The communication payloads of web services are structured text (that is, XML documents), which can be inspected, transformed, persisted, and otherwise processed with widely and even freely available tools. When efficiency demands it, however, web services also can deliver binary payloads. Finally, web services are a work in progress with real-world distributed systems as their test bed. For all of these reasons, web services are an essential tool in any modern programmer's toolbox.

The examples that follow, in this chapter and the others, are meant to be simple enough to isolate critical features of web services but also realistic enough to illustrate the power and flexibility that such services bring to software development. Let the examples begin.

A First Example

The first example is a SOAP-based web service in Java and clients in Perl, Ruby, and Java. The Java-based web service consists of an interface and an implementation.

The Service Endpoint Interface and Service Implementation Bean

The first web service in Java, like almost all of the others in this book, can be compiled and deployed using core Java SE 6 (Java Standard Edition 6) or greater without any additional software. All of the libraries required to compile, execute, and consume web services are available in core Java 6, which supports *JAX-WS* (Java API for XML-Web Services). JAX-WS supports SOAP-based and REST-style services. JAX-WS is commonly shortened to *JWS* for Java Web Services. The current version of JAX-WS is 2.x, which is a bit confusing because version 1.x has a different label: JAX-RPC. JAX-WS preserves but also significantly extends the capabilities of JAX-RPC.

A SOAP-based web service could be implemented as a single Java class but, following best practices, there should be an interface that declares the methods, which are the web service operations, and an implementation, which defines the methods declared in the interface. The interface is called the *SEI*: Service Endpoint Interface. The implementation is called the *SIB*: Service Implementation Bean. The SIB can be either a POJO or a Stateless Session *EJB* (Enterprise Java Bean). Chapter 6, which deals with the GlassFish Application Server, shows how to implement a web service as an EJB. Until then, the SOAP-based web services will be implemented as POJOs, that is, as instances of regular Java classes. These web services will be published using library classes that come with core Java 6 and, a bit later, with standalone Tomcat and GlassFish.

Core Java 6, JAX-WS, and Metro

Java SE 6 ships with JAX-WS. However, JAX-WS has a life outside of core Java 6 and a separate development team. The bleeding edge of JAX-WS is the *Metro Web Services Stack* (*https://wsit.dev.java.net*), which includes Project Tango to promote interoperability between the Java platform and *WCF* (Windows Communication Foundation), also known as Indigo. The interoperability initiative goes by the acronym *WSIT* (Web Services Interoperability Technologies). In any case, the current Metro version of JAX-WS, hereafter the *Metro release*, is typically ahead of the JAX-WS that ships with the core Java 6 SDK. With Update 4, the JAX-WS in core Java 6 went from JAX-WS 2.0 to JAX-WS 2.1, a significant improvement.

The frequent Metro releases fix bugs, add features, lighten the load on the programmer, and in general strengthen JAX-WS. At the start my goal is to introduce JAX-WS with

---

as little fuss as possible; for now, then, the JAX-WS that comes with core Java 6 is just fine. From time to time an example may involve more work than is needed under the current Metro release; in such cases, the idea is to explain what is really going on before introducing a Metro shortcut.

The Metro home page provides an easy download. Once installed, the Metro release resides in a directory named *jaxws-ri*. Subsequent examples that use the Metro release assume an environment variable `METRO_HOME`, whose value is the install directory for *jaxws-ri*. The *ri*, by the way, is short for *reference implementation*.

Finally, the downloaded Metro release is a way to do JAX-WS under core Java 5. JAX-WS requires at least core Java 5 because support for annotations begins with core Java 5.

Example 1-1 is the SEI for a web service that returns the current time as either a string or as the elapsed milliseconds from the Unix epoch, midnight January 1, 1970 GMT.

Example 1-1. Service Endpoint Interface for the TimeServer

```java
package ch01.ts;  // time server

import javax.jws.WebService;
import javax.jws.WebMethod;
import javax.jws.soap.SOAPBinding;
import javax.jws.soap.SOAPBinding.Style;

/**
 *  The annotation @WebService signals that this is the
 *  SEI (Service Endpoint Interface). @WebMethod signals
 *  that each method is a service operation.
 *
 *  The @SOAPBinding annotation impacts the under-the-hood
 *  construction of the service contract, the WSDL
 *  (Web Services Definition Language) document. Style.RPC
 *  simplifies the contract and makes deployment easier.
 */
@WebService
@SOAPBinding(style = Style.RPC) // more on this later
public interface TimeServer {
    @WebMethod String getTimeAsString();
    @WebMethod long getTimeAsElapsed();
}
```

Example 1-2 is the SIB, which implements the SEI.

Example 1-2. Service Implementation Bean for the TimeServer

```java
package ch01.ts;

import java.util.Date;
import javax.jws.WebService;
```

```
/**
 *  The @WebService property endpointInterface links the
 *  SIB (this class) to the SEI (ch01.ts.TimeServer).
 *  Note that the method implementations are not annotated
 *  as @WebMethods.
 */
@WebService(endpointInterface = "ch01.ts.TimeServer")
public class TimeServerImpl implements TimeServer {
    public String getTimeAsString() { return new Date().toString(); }
    public long getTimeAsElapsed() { return new Date().getTime(); }
}
```

The two files are compiled in the usual way from the current working directory, which in this case is immediately above the subdirectory *ch01*. The symbol % represents the command prompt:

```
% javac ch01/ts/*.java
```

A Java Application to Publish the Web Service

Once the SEI and SIB have been compiled, the web service is ready to be published. In full production mode, a Java Application Server such as BEA WebLogic, GlassFish, JBoss, or WebSphere might be used; but in development and even light production mode, a simple Java application can be used. Example 1-3 is the publisher application for the TimeServer service.

Example 1-3. Endpoint publisher for the TimeServer

```
package ch01.ts;

import javax.xml.ws.Endpoint;

/**
 * This application publishes the web service whose
 * SIB is ch01.ts.TimeServerImpl. For now, the
 * service is published at network address 127.0.0.1.,
 * which is localhost, and at port number 9876, as this
 * port is likely available on any desktop machine. The
 * publication path is /ts, an arbitrary name.
 *
 * The Endpoint class has an overloaded publish method.
 * In this two-argument version, the first argument is the
 * publication URL as a string and the second argument is
 * an instance of the service SIB, in this case
 * ch01.ts.TimeServerImpl.
 *
 * The application runs indefinitely, awaiting service requests.
 * It needs to be terminated at the command prompt with control-C
 * or the equivalent.
 *
 * Once the applicatation is started, open a browser to the URL
 *
```

```
*      http://127.0.0.1:9876/ts?wsdl
*
* to view the service contract, the WSDL document. This is an
* easy test to determine whether the service has deployed
* successfully. If the test succeeds, a client then can be
* executed against the service.
*/
public class TimeServerPublisher {
    public static void main(String[ ] args) {
        // 1st argument is the publication URL
        // 2nd argument is an SIB instance
        Endpoint.publish("http://127.0.0.1:9876/ts", new TimeServerImpl());
    }
}
```

Once compiled, the publisher can be executed in the usual way:

```
% java ch01.ts.TimeServerPublisher
```

How the Endpoint Publisher Handles Requests

Out of the box, the Endpoint publisher handles one client request at a time. This is fine for getting web services up and running in development mode. However, if the processing of a given request should hang, then all other client requests are effectively blocked. An example at the end of this chapter shows how Endpoint can handle requests concurrently so that one hung request does not block the others.

Testing the Web Service with a Browser

We can test the deployed service by opening a browser and viewing the *WSDL* (Web Service Definition Language) document, which is an automatically generated service contract. (WSDL is pronounced "whiz dull.") The browser is opened to a URL that has two parts. The first part is the URL published in the Java TimeServerPublisher application: *http://127.0.0.1:9876/ts*. Appended to this URL is the query string *?wsdl* in upper-, lower-, or mixed case. The result is *http://127.0.0.1:9876/ts?wsdl*. Example 1-4 is the WSDL document that the browser displays.

Example 1-4. WSDL document for the TimeServer service

```
<?xml version="1.0" encoding="UTF-8"?>
<definitions
    xmlns="http://schemas.xmlsoap.org/wsdl/"
    xmlns:tns="http://ts.ch01/"
    xmlns:xsd="http://www.w3.org/2001/XMLSchema"
    xmlns:soap="http://schemas.xmlsoap.org/wsdl/soap/"
    targetNamespace="http://ts.ch01/"
    name="TimeServerImplService">
  <types></types>

  <message name="getTimeAsString"></message>
  <message name="getTimeAsStringResponse">
```

```
    <part name="return" type="xsd:string"></part>
  </message>
  <message name="getTimeAsElapsed"></message>
  <message name="getTimeAsElapsedResponse">
    <part name="return" type="xsd:long"></part>
  </message>

  <portType name="TimeServer">
    <operation name="getTimeAsString" parameterOrder="">
      <input message="tns:getTimeAsString"></input>
      <output message="tns:getTimeAsStringResponse"></output>
    </operation>
    <operation name="getTimeAsElapsed" parameterOrder="">
      <input message="tns:getTimeAsElapsed"></input>
      <output message="tns:getTimeAsElapsedResponse"></output>
    </operation>
  </portType>

  <binding name="TimeServerImplPortBinding" type="tns:TimeServer">
    <soap:binding style="rpc"
                  transport="http://schemas.xmlsoap.org/soap/http">
    </soap:binding>
    <operation name="getTimeAsString">
      <soap:operation soapAction=""></soap:operation>
      <input>
        <soap:body use="literal" namespace="http://ts.ch01/"></soap:body>
      </input>
      <output>
        <soap:body use="literal" namespace="http://ts.ch01/"></soap:body>
      </output>
    </operation>
    <operation name="getTimeAsElapsed">
      <soap:operation soapAction=""></soap:operation>
      <input>
        <soap:body use="literal" namespace="http://ts.ch01/"></soap:body>
      </input>
      <output>
        <soap:body use="literal" namespace="http://ts.ch01/"></soap:body>
      </output>
    </operation>
  </binding>

  <service name="TimeServerImplService">
    <port name="TimeServerImplPort" binding="tns:TimeServerImplPortBinding">
      <soap:address location="http://localhost:9876/ts"></soap:address>
    </port>
  </service>
</definitions>
```

Chapter 2 examines the WSDL in detail and introduces Java utilities associated with
the service contract. For now, two sections of the WSDL (both shown in bold) deserve
a quick look. The portType section, near the top, groups the operations that the web
service delivers, in this case the operations getTimeAsString and getTimeAsElapsed,
which are the two Java methods declared in the SEI and implemented in the SIB. The

WSDL `portType` is like a Java interface in that the `portType` presents the service operations abstractly but provides no implementation detail. Each operation in the web service consists of an `input` and an `output` message, where *input* means *input for the web service*. At runtime, each message is a SOAP document. The other WSDL section of interest is the last, the `service` section, and in particular the service `location`, in this case the URL *http://localhost:9876/ts*. The URL is called the *service endpoint* and it informs clients about where the service can be accessed.

The WSDL document is useful for both creating and executing clients against a web service. Various languages have utilities for generating client-support code from a WSDL. The core Java utility is now called *wsimport* but the earlier names *wsdl2java* and *java2wsdl* were more descriptive. At runtime, a client can consume the WSDL document associated with a web service in order to get critical information about the data types associated with the operations bundled in the service. For example, a client could determine from our first WSDL that the operation `getTimeAsElapsed` returns an integer and expects no arguments.

The WSDL also can be accessed with various utilities such as *curl*. For example, the command:

```
% curl http://localhost:9876/ts?wsdl
```

also displays the WSDL.

Avoiding a Subtle Problem in the Web Service Implementation

This example departs from the all-too-common practice of having the service's SIB (the class `TimeServerImpl`) connected to the SEI (the interface `TimeServer`) only through the `endpointInterface` attribute in the `@WebService` annotation. It is not unusual to see this:

```
@WebService(endpointInterface = "ch01.ts.TimeServer")
public class TimeServerImpl { // implements TimeServer removed
```

The style is popular but unsafe. It is far better to have the `implements` clause so that the compiler checks whether the SIB implements the methods declared in the SEI. Remove the `implements` clause and comment out the definitions of the two web service operations:

```
@WebService(endpointInterface = "ch01.ts.TimeServer")
public class TimeServerImpl {
    // public String getTimeAsString() { return new Date().toString(); }
    // public long gettimeAsElapsed() { return new Date().getTime(); }
}
```

The code still compiles. With the `implements` clause in place, the compiler issues a fatal error because the SIB fails to define the methods declared in the SEI.

A Perl and a Ruby Requester of the Web Service

To illustrate the language transparency of web services, the first client against the Java-based web service is not in Java but rather in Perl. The second client is in Ruby. Example 1-5 is the Perl client.

Example 1-5. Perl client for the TimeServer client

```
#!/usr/bin/perl -w

use SOAP::Lite;
my $url = 'http://127.0.0.1:9876/ts?wsdl';
my $service = SOAP::Lite->service($url);

print "\nCurrent time is: ", $service->getTimeAsString();
print "\nElapsed milliseconds from the epoch: ", $service->getTimeAsElapsed();
```

On a sample run, the output was:

```
Current time is: Thu Oct 16 21:37:35 CDT 2008
Elapsed milliseconds from the epoch: 1224211055700
```

The Perl module `SOAP::Lite` provides the under-the-hood functionality that allows the client to issue the appropriate SOAP request and to process the resulting SOAP response. The request URL, the same URL used to test the web service in the browser, ends with a query string that asks for the WSDL document. The Perl client gets the WSDL document from which the `SOAP::Lite` library then generates the appropriate service object (in Perl syntax, the scalar variable `$service`). By consuming the WSDL document, the `SOAP::Lite` library gets the information needed, in particular, the names of the web service operations and the data types involved in these operations. Figure 1-2 depicts the architecture.

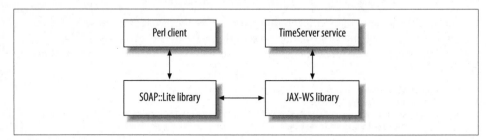

Figure 1-2. Architecture of the Perl client and Java service

After the setup, the Perl client invokes the web service operations without any fuss. The SOAP messages remain unseen.

Example 1-6 is a Ruby client that is functionally equivalent to the Perl client.

Example 1-6. Ruby client for the TimeServer client

```ruby
#!/usr/bin/ruby

# one Ruby package for SOAP-based services
require 'soap/wsdlDriver'

wsdl_url = 'http://127.0.0.1:9876/ts?wsdl'


service = SOAP::WSDLDriverFactory.new(wsdl_url).create_rpc_driver

# Save request/response messages in files named '...soapmsgs...'
service.wiredump_file_base = 'soapmsgs'

# Invoke service operations.
result1 = service.getTimeAsString
result2 = service.getTimeAsElapsed

# Output results.
puts "Current time is: #{result1}"
puts "Elapsed milliseconds from the epoch: #{result2}"
```

The Hidden SOAP

In SOAP-based web services, a client typically makes a remote procedure call against the service by invoking one of the web service operations. As mentioned earlier, this back and forth between the client and service is the request/response message exchange pattern, and the SOAP messages exchanged in this pattern allow the web service and a consumer to be programmed in different languages. We now look more closely at what happens under the hood in our first example. The Perl client generates an HTTP request, which is itself a formatted message whose *body* is a SOAP message. Example 1-7 is the HTTP request from a sample run.

Example 1-7. HTTP request for the TimeServer service

```
POST http://127.0.0.1:9876/ts HTTP/ 1.1
Accept: text/xml
Accept: multipart/*
Accept: application/soap
User-Agent: SOAP::Lite/Perl/0.69
Content-Length: 434
Content-Type: text/xml; charset=utf-8
SOAPAction: ""

<?xml version="1.0" encoding="UTF-8"?>
<soap:Envelope
    soap:encodingStyle="http:// schemas.xmlsoap.org/soap/encoding/"
    xmlns:soap="http://schemas.xmlsoap.org/soap/ envelope/"
    xmlns:soapenc="http://schemas.xmlsoap.org/soap/encoding/"
    xmlns:xsi="http://www.w3.org/2001/XMLSchema-instance"
    xmlns:tns="http://ts.ch01/"
```

```
    xmlns:xsd ="http://www.w3.org/2001/XMLSchema">
  <soap:Body>
    <tns:getTimeAsString xsi:nil="true" />
  </soap:Body>
</soap:Envelope>
```

The HTTP request is a message with a structure of its own. In particular:

- The HTTP *start line* comes first and specifies the request method, in this case the POST method, which is typical of requests for dynamic resources such as web services or other web application code (for example, a Java servlet) as opposed to requests for a static HTML page. In this case, a POST rather than a GET request is needed because only a POST request has a body, which encapsulates the SOAP message. Next comes the request URL followed by the HTTP version, in this case 1.1, that the requester understands. HTTP 1.1 is the current version.

- Next come the HTTP *headers*, which are key/value pairs in which a colon (:) separates the key from the value. The order of the key/value pairs is arbitrary. The key `Accept` occurs three times, with a *MIME* (Multipurpose Internet Mail Extensions) type/subtype as the value: `text/xml`, `multipart/*`, and `application/soap`. These three pairs signal that the requester is ready to accept an arbitrary XML response, a response with arbitrarily many attachments of any type (a SOAP message can have arbitrarily many attachments), and a SOAP document, respectively. The HTTP key `SOAPAction` is often present in the HTTP header of a web service request and the key's value may be the empty string, as in this case; but the value also might be the name of the requested web service operation.

- Two *CRLF* (Carriage Return Line Feed) characters, which correspond to two Java \n characters, separate the HTTP headers from the HTTP body, which is required for the POST verb but may be empty. In this case, the HTTP body contains the SOAP document, commonly called the SOAP envelope because the outermost or *document* element is named `Envelope`. In this SOAP envelope, the SOAP body contains a single element whose *local name* is getTimeAsString, which is the name of the web service operation that the client wants to invoke. The SOAP request envelope is simple in this example because the requested operation takes no arguments.

On the web service side, the underlying Java libraries process the HTTP request, extract the SOAP envelope, determine the identity of the requested service operation, invoke the corresponding Java method `getTimeAsString`, and then generate the appropriate SOAP message to carry the method's return value back to the client. Example 1-8 is the HTTP response from the Java `TimeServerImpl` service request shown in Example 1-7.

Example 1-8. HTTP response from the TimeServer service

```
HTTP/1.1 200 OK
Content-Length: 323
Content-Type: text/xml; charset=utf-8
```

```
Client-Date: Mon, 28 Apr 2008 02:12:54 GMT
Client-Peer: 127.0.0.1:9876
Client-Response-Num: 1

<?xml version="1.0" ?>
<soapenv:Envelope
    xmlns:soapenv="http://schemas.xmlsoap.org/soap/envelope/"
    xmlns:xsd="http://www.w3.org/2001/XMLSchema">
  <soapenv:Body>
    <ans:getTimeAsStringResponse xmlns:ans="http://ts.ch01/">
      <return>Mon Apr 28 14:12:54 CST 2008</return>
    </ans:getTimeAsStringResponse>
  </soapenv:Body>
</soapenv:Envelope>
```

Once again the SOAP envelope is the body of an HTTP message, in this case the HTTP response to the client. The HTTP *start line* now contains the status code as the integer 200 and the corresponding text OK, which signal that the client request was handled successfully. The SOAP envelope in the HTTP response's body contains the current time as a string between the XML start and end tags named **return**. The Perl SOAP library extracts the SOAP envelope from the HTTP response and, because of information in the WSDL document, expects the desired return value from the web service operation to occur in the XML **return** element.

A Java Requester of the Web Service

Example 1-9 is a Java client functionally equivalent to the Perl and Ruby clients shown in Examples 1-5 and 1-6, respectively.

Example 1-9. Java client for the Java web service

```java
package ch01.ts;

import javax.xml.namespace.QName;
import javax.xml.ws.Service;
import java.net.URL;
class TimeClient {
    public static void main(String args[ ]) throws Exception {
        URL url = new URL("http://localhost:9876/ts?wsdl");

        // Qualified name of the service:
        //    1st arg is the service URI
        //    2nd is the service name published in the WSDL
        QName qname = new QName("http://ts.ch01/", "TimeServerImplService");

        // Create, in effect, a factory for the service.
        Service service = Service.create(url, qname);

        // Extract the endpoint interface, the service "port".
        TimeServer eif = service.getPort(TimeServer.class);
```

```
        System.out.println(eif.getTimeAsString());
        System.out.println(eif.getTimeAsElapsed());
    }
}
```

The Java client uses the same URL with a query string as do the Perl and Ruby clients, but the Java client explicitly creates an XML *qualified name*, which has the syntax *namespace URI:local name*. A *URI* is a Uniform Resource Identifier and differs from the more common URL in that a URL specifies a *location*, whereas a URI need not specify a location. In short, a URI need not be a URL. For now, it is enough to underscore that the Java class `java.xml.namespace.QName` represents an XML-qualified name. In this example, the namespace URI is provided in the WSDL, and the local name is the SIB class name `TimeServerImpl` with the word `Service` appended. The local name occurs in the `service` section, the last section of the WSDL document.

Once the `URL` and `QName` objects have been constructed and the `Service.create` method has been invoked, the statement of interest:

```
    TimeServer eif = service.getPort(TimeServer.class);
```

executes. Recall that, in the WSDL document, the `portType` section describes, in the style of an interface, the operations included in the web service. The `getPort` method returns a reference to a Java object that can invoke the `portType` operations. The `eif` object reference is of type `ch01.ts.TimeServer`, which is the SEI type. The Java client, like the Perl client, invokes the two web service methods; and the Java libraries, like the Perl and Ruby libraries, generate and process the SOAP messages exchanged transparently to enable the successful method invocations.

Wire-Level Tracking of HTTP and SOAP Messages

Example 1-7 and Example 1-8 show an HTTP request message and an HTTP response message, respectively. Each HTTP message encapsulates a SOAP envelope. These message traces were done with the Perl client by changing the Perl `use` directive in Example 1-5:

```
    use SOAP::Lite;
```

to:

```
    use SOAP::Lite +trace;
```

The Ruby client in Example 1-6 contains a line:

```
    service.wiredump_file_base = 'soapmsgs'
```

that causes the SOAP envelopes to be saved in files on the local disk. It is possible to capture the wire-level traffic directly in Java as well, as later examples illustrate. Various options are available for tracking SOAP and HTTP messages at the wire level. Here is a short introduction to some of them.

The *tcpmon* utility (available at *https://tcpmon.dev.java.net*) is free and downloads as an executable JAR file. Its graphical user interface (GUI) is easy to use. The utility requires only three settings: the server's name, which defaults to `localhost`; the server's port, which would be set to `9876` for the `TimeServer` example because this is the port at which `Endpoint` publishes the service; and the local port, which defaults to `8080` and is the port at which *tcpmon* listens. With *tcpmon* in use, the `TimeClient` would send its requests to port `8080` instead of port `9876`. The *tcpmon* utility intercepts HTTP traffic between the client and web service, displaying the full messages in its GUI.

The Metro release has utility classes for tracking HTTP and SOAP traffic. This approach does not require any change to the client or to the service code; however, an additional package must be put on the classpath and a system property must be set either at the command line or in code. The required package is in the file *jaxws_ri/jaxws-rt.jar*. Assuming that the environment variable `METRO_HOME` points to the *jaxws-ri* directory, here is the command that tracks HTTP and SOAP traffic between the `TimeClient`, which connects to the service on port 9876, and the `TimeServer` service. (Under Windows, `$METRO_HOME` becomes `%METRO_HOME%`.) The command is on three lines for readability:

```
% java -cp ".":$METRO_HOME/lib/jaxws-rt.jar \
  -Dcom.sun.xml.ws.transport.http.client.HttpTransportPipe.dump=true \
  ch01.ts.TimeClient
```

The resulting dump shows all of the SOAP traffic but not all of the HTTP headers. Message tracking also can be done on the service side.

There are various other open source and commercial products available for tracking the SOAP traffic. Among the products that are worth a look at are SOAPscope (*http://home.mindreef.com*), NetSniffer (*http://www.miray.de*), and Wireshark (*http://www.wireshark.org*). The *tcpdump* utility comes with most Unix-type systems, including Linux and OS X, and is available on Windows as WinDump (*http://www.winpcap.org*). Besides being free, *tcpdump* is nonintrusive in that it requires no change to either a web service or a client. The *tcpdump* utility dumps message traffic to the standard output. The companion utility *tcptrace* (*http://www.tcptrace.org*) can be used to analyze the dump. The remainder of this section briefly covers *tcpdump* as a flexible and powerful trace utility.

Under Unix-type systems, the *tcpdump* utility typically must be executed as *superuser*. There are various flagged arguments that determine how the utility works. Here is a sample invocation:

```
% tcpdump -i lo -A -s 1024 -l 'dst host localhost and port 9876' | tee dump.log
```

The utility can capture packets on any network interface. A list of such interfaces is available with the *tcpdump -D* (under Windows, *WinDump -D*), which is equivalent to the *ifconfig -a* command on Unix-like systems. In this example, the flag/value pair `-i lo` means *capture packets from the interface lo*, where *lo* is short for the *localhost* network interface on many Unix-like systems. The flag `-A` means that the captured packets should be presented in ASCII, which is useful for web packets as these typically contain text. The `-s 1024` flag sets the *snap length*, the number of bytes that should be captured from each packet. The flag `-1` forces the standard output to be line buffered and easier to read; and, on the same theme, the construct | `tee dump.log` at the end pipes the same output that shows up on the screen (the standard output) into a local file named *dump.log*. Finally, the expression:

```
'dst host localhost and port 9876'
```

acts as a filter, capturing only packets whose destination is *localhost* on port 9876, the port on which `TimeServerPublisher` of Example 1-3 publishes the `TimeServer` service.

The *tcpdump* utility and the `TimeServerPublisher` application can be started in any order. Once both are running, the `TimeClient` or one of the other clients can be executed. With the sample use of *tcpdump* shown above, the underlying network packets are saved in the file *dump.log*. The file does require some editing to make it easily readable. In any case, the *dump.log* file captures the same SOAP envelopes shown in Examples 1-7 and 1-8.

What's Clear So Far?

The first example is a web service with two operations, each of which delivers the current time but in different representations: in one case as a human-readable string, and in the other case as the elapsed milliseconds from the Unix epoch. The two operations are implemented as independent, self-contained methods. From the service requester's perspective, either method may be invoked independently of the other and one invocation of a service method has no impact on any subsequent invocation of the same service method. The two Java methods depend neither on one another nor on any instance field to which both have access; indeed, the SIB class `TimeServerImpl` has no fields at all. In short, the two method invocations are stateless.

In the first example, neither method expects arguments. In general, web service operations may be parameterized so that appropriate information can be passed to the operation as part of the service request. Regardless of whether the web service operations are parameterized, they still should appear to the requester as independent and self-contained. This design principle will guide all of the samples that we consider, even ones that are richer than the first.

Key Features of the First Code Example

The `TimeServerImpl` class implements a web service with a distinctive message exchange pattern (MEP)—request/response. The service allows a client to make a language-neutral remote procedure call, invoking the methods `getTimeAsString` and `getTimeAsElapsed`. Other message patterns are possible. Imagine, for example, a web service that tracks new snow amounts for ski areas. Some participating clients, perhaps snow-measuring electrical devices strategically placed around the ski slopes, might use the one-way pattern by sending a snow amount from a particular location but without expecting a response from the service. The service might exhibit the notification pattern by multicasting to subscribing clients (for instance, travel bureaus) information about current snow conditions. Finally, the service might periodically use the solicit/response pattern to ask a subscribing client whether the client wishes to continue receiving notifications. In summary, SOAP-based web services support various patterns. The request/response pattern of RPC remains the dominant one. The infrastructure needed to support this pattern in particular is worth summarizing:

Message transport

SOAP is designed to be transport-neutral, a design goal that complicates matters because SOAP messages cannot rely on protocol-specific information included in the transport infrastructure. For instance, SOAP delivered over HTTP should not differ from SOAP delivered over some other transport protocol such as *SMTP* (Simple Mail Transfer Protocol), *FTP* (File Transfer Protocol), or even *JMS* (Java Message Service). In practice, however, HTTP is the usual transport for SOAP-based services, a point underscored in the usual name: SOAP-based *web* services.

Service contract

The service client requires information about the service's operations in order to invoke them. In particular, the client needs information about the invocation syntax: the operation's name, the order and types of the arguments passed to the operation, and the type of the returned value. The client also requires the service endpoint, typically the service URL. The WSDL document provides these pieces of information and others. Although a client could invoke a service without first accessing the WSDL, this would make things harder than they need to be.

Type system

The key to language neutrality and, therefore, service/consumer interoperability is a shared type system so that the data types used in the client's invocation coordinate with the types used in the service operation. Consider a simple example. Suppose that a Java web service has the operation:

```
boolean bytes_ok(byte[ ] some_bytes)
```

The bytes_ok operation performs some validation test on the bytes passed to the operation as an array argument, returning either true or false. Now assume that a client written in C needs to invoke bytes_ok. C has no types named boolean and byte. C represents boolean values with integers, with nonzero as true and zero as false; and the C type signed char corresponds to the Java type byte. A web service would be cumbersome to consume if clients had to map client-language types to service-language types. In SOAP-based web services, the XML Schema type system is the default type system that mediates between the client's types and the service's types. In the example above, the XML Schema type xsd:byte is the type that mediates between the C signed char and the Java byte; and the XML Schema type xsd:boolean is the mediating type for the C integers nonzero and zero and the Java boolean values true and false. In the notation xsd:byte, the prefix xsd (XML Schema Definition) underscores that this is an XML Schema type because xsd is the usual extension for a file that contains an XML Schema definition; for instance, *purchaseOrder.xsd*.

Java's SOAP API

A major appeal of SOAP-based web services is that the SOAP usually remains hidden. Nonetheless, it may be useful to glance at Java's underlying support for generating and processing SOAP messages. Chapter 3, which introduces SOAP handlers, puts the SOAP API to practical use. This section provides a first look at the SOAP API through a simulation example. The application consists of one class, `DemoSoap`, but simulates sending a SOAP message as a request and receiving another as a response. Example 1-10 shows the full application.

Example 1-10. A demonstration of Java's SOAP API

```
package ch01.soap;

import java.util.Date;
import java.util.Iterator;
import java.io.InputStream;
import java.io.ByteArrayOutputStream;
import java.io.ByteArrayInputStream;
import java.io.IOException;
import javax.xml.soap.MessageFactory;
import javax.xml.soap.SOAPMessage;
import javax.xml.soap.SOAPEnvelope;
import javax.xml.soap.SOAPHeader;
import javax.xml.soap.SOAPBody;
import javax.xml.soap.SOAPPart;
import javax.xml.soap.SOAPElement;
import javax.xml.soap.SOAPException;
import javax.xml.soap.Node;
import javax.xml.soap.Name;

public class DemoSoap {
    private static final String LocalName = "TimeRequest";
    private static final String Namespace = "http://ch01/mysoap/";
    private static final String NamespacePrefix = "ms";

    private ByteArrayOutputStream out;
    private ByteArrayInputStream in;

    public static void main(String[ ] args) {
        new DemoSoap().request();
    }

    private void request() {
        try {
            // Build a SOAP message to send to an output stream.
            SOAPMessage msg = create_soap_message();
```

```
   // Inject the appropriate information into the message.
   // In this case, only the (optional) message header is used
   // and the body is empty.
   SOAPEnvelope env = msg.getSOAPPart().getEnvelope();
   SOAPHeader hdr = env.getHeader();

   // Add an element to the SOAP header.
   Name lookup_name = create_qname(msg);
   hdr.addHeaderElement(lookup_name).addTextNode("time_request");

   // Simulate sending the SOAP message to a remote system by
   // writing it to a ByteArrayOutputStream.
   out = new ByteArrayOutputStream();
   msg.writeTo(out);

   trace("The sent SOAP message:", msg);

   SOAPMessage response = process_request();
   extract_contents_and_print(response);
   }
   catch(SOAPException e) { System.err.println(e); }
   catch(IOException e) { System.err.println(e); }
}

private SOAPMessage process_request() {
   process_incoming_soap();
   coordinate_streams();
   return create_soap_message(in);
}

private void process_incoming_soap() {
   try {
      // Copy output stream to input stream to simulate
      // coordinated streams over a network connection.
      coordinate_streams();

      // Create the "received" SOAP message from the
      // input stream.
      SOAPMessage msg = create_soap_message(in);

      // Inspect the SOAP header for the keyword 'time_request'
      // and process the request if the keyword occurs.
      Name lookup_name = create_qname(msg);

      SOAPHeader header = msg.getSOAPHeader();
      Iterator it = header.getChildElements(lookup_name);
      Node next = (Node) it.next();
      String value = (next == null) ? "Error!" : next.getValue();

      // If SOAP message contains request for the time, create a
      // new SOAP message with the current time in the body.
      if (value.toLowerCase().contains("time_request")) {
```

```java
      // Extract the body and add the current time as an element.
      String now = new Date().toString();
      SOAPBody body = msg.getSOAPBody();
      body.addBodyElement(lookup_name).addTextNode(now);
      msg.saveChanges();

      // Write to the output stream.
      msg.writeTo(out);
      trace("The received/processed SOAP message:", msg);
    }
  }
  catch(SOAPException e) { System.err.println(e); }
  catch(IOException e) { System.err.println(e); }
}

private void extract_contents_and_print(SOAPMessage msg) {
  try {
    SOAPBody body = msg.getSOAPBody();

    Name lookup_name = create_qname(msg);
    Iterator it = body.getChildElements(lookup_name);
    Node next = (Node) it.next();

    String value = (next == null) ? "Error!" : next.getValue();
    System.out.println("\n\nReturned from server: " + value);
  }
  catch(SOAPException e) { System.err.println(e); }
}

private SOAPMessage create_soap_message() {
  SOAPMessage msg = null;
  try {
    MessageFactory mf = MessageFactory.newInstance();
    msg = mf.createMessage();
  }
  catch(SOAPException e) { System.err.println(e); }
  return msg;
}

private SOAPMessage create_soap_message(InputStream in) {
  SOAPMessage msg = null;
  try {
    MessageFactory mf = MessageFactory.newInstance();
    msg = mf.createMessage(null, // ignore MIME headers
                           in);  // stream source
  }
  catch(SOAPException e) { System.err.println(e); }
  catch(IOException e) { System.err.println(e); }
  return msg;
}

private Name create_qname(SOAPMessage msg) {
  Name name = null;
```

```
    try {
      SOAPEnvelope env = msg.getSOAPPart().getEnvelope();
      name = env.createName(LocalName, NamespacePrefix, Namespace);
    }
    catch(SOAPException e) { System.err.println(e); }
    return name;
  }

  private void trace(String s, SOAPMessage m) {
    System.out.println("\n");
    System.out.println(s);
    try {
      m.writeTo(System.out);
    }
    catch(SOAPException e) { System.err.println(e); }
    catch(IOException e) { System.err.println(e); }
  }

  private void coordinate_streams() {
    in = new ByteArrayInputStream(out.toByteArray());
    out.reset();
  }
}
```

Here is a summary of how the application runs, with emphasis on the code involving SOAP messages. The DemoSoap application's request method generates a SOAP message and adds the string time_request to the SOAP envelope's header. The code segment, with comments removed, is:

```
SOAPMessage msg = create_soap_message();
SOAPEnvelope env = msg.getSOAPPart().getEnvelope();
SOAPHeader hdr = env.getHeader();
Name lookup_name = create_qname(msg);
hdr.addHeaderElement(lookup_name).addTextNode("time_request");
```

There are two basic ways to create a SOAP message. The simple way is illustrated in this code segment:

```
MessageFactory mf = MessageFactory.newInstance();
SOAPMessage msg = mf.createMessage();
```

In the more complicated way, the MessageFactory code is the same, but the creation call becomes:

```
SOAPMessage msg = mf.createMessage(mime_headers, input_stream);
```

The first argument to createMessage is a collection of the transport-layer headers (for instance, the key/value pairs that make up an HTTP header), and the second argument is an input stream that provides the bytes to create the message (for instance, the input stream encapsulated in a Java Socket instance).

Once the SOAP message is created, the header is extracted from the SOAP envelope and an XML text node is inserted with the value `time_request`. The resulting SOAP message is:

```
<SOAP-ENV:Envelope
    xmlns:SOAP-ENV="http://schemas.xmlsoap.org/soap/envelope/">
    <SOAP-ENV:Header>
      <ms:TimeRequest xmlns:ms="http://ch01/mysoap/">
        time_request
      </ms:TimeRequest>
    </SOAP-ENV:Header>
    <SOAP-ENV:Body/>
</SOAP-ENV:Envelope>
```

There is no need right now to examine every detail of this SOAP message. Here is a summary of some key points. The SOAP body is always required but, as in this case, the body may be empty. The SOAP header is optional but, in this case, the header contains the text `time_request`. Message contents such as `time_request` normally would be placed in the SOAP body and special processing information (for instance, user authentication data) would be placed in the header. The point here is to illustrate how the SOAP header and the SOAP body can be manipulated.

The `request` method writes the SOAP message to a `ByteArrayOutputStream`, which simulates sending the message over a network connection to a receiver on a different host. The `request` method invokes the `process_request` method, which in turn delegates the remaining tasks to other methods. The processing goes as follows. The received SOAP message is created from a `ByteArrayInputStream`, which simulates an input stream on the receiver's side; this stream contains the sent SOAP message. The SOAP message now is created from the input stream:

```
SOAPMessage msg = null;
try {
    MessageFactory mf = MessageFactory.newInstance();
    msg = mf.createMessage(null, // ignore MIME headers
                           in);  // stream source (ByteArrayInputStream)
}
```

and then the SOAP message is processed to extract the `time_request` string. The extraction goes as follows. First, the SOAP header is extracted from the SOAP message and an iterator over the elements with the tag name:

```
<ms:TimeRequest xmlns:ms="http://ch01/mysoap/>
```

is created. In this example, there is one element with this tag name and the element should contain the string `time_request`. The lookup code is:

```
SOAPHeader header = msg.getSOAPHeader();
Iterator it = header.getChildElements(lookup_name);
Node next = (Node) it.next();
String value = (next == null) ? "Error!" : next.getValue();
```

If the SOAP header contains the proper request string, the SOAP body is extracted from the incoming SOAP message and an element containing the current time as a string is added to the SOAP body. The revised SOAP message then is sent as a response. Here is the code segment with the comments removed:

```
if (value.toLowerCase().contains("time_request")) {
    String now = new Date().toString();
    SOAPBody body = msg.getSOAPBody();
    body.addBodyElement(lookup_name).addTextNode(now);
    msg.saveChanges();

    msg.writeTo(out);
    trace("The received/processed SOAP message:", msg);
}
```

The outgoing SOAP message on a sample run was:

```
<SOAP-ENV:Envelope
    xmlns:SOAP-ENV="http://schemas.xmlsoap.org/soap/envelope/">
  <SOAP-ENV:Header>
      <ms:TimeRequest xmlns:ms="http://ch01/mysoap/">
          time_request
      </ms:TimeRequest>
  </SOAP-ENV:Header>
  <SOAP-ENV:Body>
      <ms:TimeRequest xmlns:ms="http://ch01/mysoap/">
          Mon Oct 27 14:45:53 CDT 2008
      </ms:TimeRequest>
  </SOAP-ENV:Body>
</SOAP-ENV:Envelope>
```

This example provides a first look at Java's API. Later examples illustrate production-level use of the SOAP API.

An Example with Richer Data Types

The operations in the TimeServer service take no arguments and return simple types, a string and an integer. This section offers a richer example whose details are clarified in the next chapter.

The Teams web service in Example 1-11 differs from the TimeServer service in several important ways.

Example 1-11. The Teams document-style web service

```
package ch01.team;

import java.util.List;
import javax.jws.WebService;
import javax.jws.WebMethod;

@WebService
public class Teams {
    private TeamsUtility utils;
```

```
    public Teams() {
        utils = new TeamsUtility();
        utils.make_test_teams();
    }

    @WebMethod
    public Team getTeam(String name) { return utils.getTeam(name); }

    @WebMethod
    public List<Team> getTeams() { return utils.getTeams(); }
}
```

For one, the Teams service is implemented as a single Java class rather than as a separate
SEI and SIB. This is done simply to illustrate the possibility. A more important difference is in the return types of the two Teams operations. The operation getTeam is parameterized and returns an object of the programmer-defined type Team, which is a list
of Player instances, another programmer-defined type. The operation getTeams returns
a List<Team>, that is, a Java Collection.

The utility class TeamsUtility generates the data. In a production environment, this
utility might retrieve a team or list of teams from a database. To keep this example
simple, the utility instead creates the teams and their players on the fly. Here is part of
the utility:

```
    package ch01.team;

    import java.util.Set;
    import java.util.List;
    import java.util.ArrayList;
    import java.util.Map;
    import java.util.HashMap;

    public class TeamsUtility {
        private Map<String, Team> team_map;

        public TeamsUtility() {
            team_map = new HashMap<String, Team>();
        }

        public Team getTeam(String name) { return team_map.get(name); }
        public List<Team> getTeams() {
            List<Team> list = new ArrayList<Team>();
            Set<String> keys = team_map.keySet();
            for (String key : keys)
                list.add(team_map.get(key));
            return list;
        }

        public void make_test_teams() {
            List<Team> teams = new ArrayList<Team>();
            ...
            Player chico = new Player("Leonard Marx", "Chico");
            Player groucho = new Player("Julius Marx", "Groucho");
```

```
    Player harpo = new Player("Adolph Marx", "Harpo");
    List<Player> mb = new ArrayList<Player>();
    mb.add(chico); mb.add(groucho); mb.add(harpo);
    Team marx_brothers = new Team("Marx Brothers", mb);
    teams.add(marx_brothers);

    store_teams(teams);

  }

  private void store_teams(List<Team> teams) {
    for (Team team : teams)
      team_map.put(team.getName(), team);
  }
}
```

Publishing the Service and Writing a Client

Recall that the SEI for the `TimeServer` service contains the annotation:

```
@SOAPBinding(style = Style.RPC)
```

This annotation requires that the service use only very simple types such as string and integer. By contrast, the `Teams` service uses richer data types, which means that `Style.DOCUMENT`, the default, should replace `Style.RPC`. The document style does require more setup, which is given below but not explained until the next chapter. Here, then, are the steps required to get the web service deployed and a sample client written quickly:

1. The source files are compiled in the usual way. From the working directory, which has *ch01* as a subdirectory, the command is:

    ```
    % javac ch01/team/*.java
    ```

    In addition to the `@WebService`-annotated `Teams` class, the *ch01/team* directory contains the `Team`, `Player`, `TeamsUtility`, and `TeamsPublisher` classes shown below all together:

    ```
    package ch01.team;
    public class Player {
        private String name;
        private String nickname;

        public Player() { }
        public Player(String name, String nickname) {
            setName(name);
            setNickname(nickname);
        }

        public void setName(String name) { this.name = name; }
        public String getName() { return name; }
        public void setNickname(String nickname) { this.nickname = nickname; }
        public String getNickname() { return nickname; }
    }
    ```

```
// end of Player.java

package ch01.team;

import java.util.List;
public class Team {
    private List<Player> players;
    private String name;

    public Team() { }
    public Team(String name, List<Player> players) {
        setName(name);
        setPlayers(players);
    }

        public void setName(String name) { this.name = name; }
        public String getName() { return name; }
        public void setPlayers(List<Player> players) { this.players = players; }
        public List<Player> getPlayers() { return players; }
        public void setRosterCount(int n) { } // no-op but needed for property
        public int getRosterCount() { return (players == null) ? 0 : players.size(); }

    }
}
// end of Team.java

package ch01.team;
import javax.xml.ws.Endpoint;
class TeamsPublisher {
    public static void main(String[ ] args) {
        int port = 8888;
        String url = "http://localhost:" + port + "/teams";
        System.out.println("Publishing Teams on port " + port);
        Endpoint.publish(url, new Teams());
    }
}
```

2. In the working directory, invoke the *wsgen* utility, which comes with core Java 6:

```
% wsgen -cp . ch01.team.Teams
```

This utility generates various *artifacts*; that is, Java types needed by the method `Endpoint.publish` to generate the service's WSDL. Chapter 2 looks closely at these artifacts and how they contribute to the WSDL.

3. Execute the `TeamsPublisher` application.

4. In the working directory, invoke the *wsimport* utility, which likewise comes with core Java 6:

```
% wsimport -p teamsC -keep http://localhost:8888/teams?wsdl
```

This utility generates various classes in the subdirectory *teamsC* (the -p flag stands for `package`). These classes make it easier to write a client against the service.

Step 4 expedites the coding of a client, which is shown here:

```
import teamsC.TeamsService;
import teamsC.Teams;
import teamsC.Team;
import teamsC.Player;
import java.util.List;

class TeamClient {
    public static void main(String[ ] args) {
        TeamsService service = new TeamsService();
        Teams port = service.getTeamsPort();
        List<Team> teams = port.getTeams();
        for (Team team : teams) {
            System.out.println("Team name: " + team.getName() +
                                " (roster count: " + team.getRosterCount() + ")");
            for (Player player : team.getPlayers())
                System.out.println("  Player: " + player.getNickname());
        }
    }
}
```

When the client executes, the output is:

```
Team name: Abbott and Costello (roster count: 2)
  Player: Bud
  Player: Lou
Team name: Marx Brothers (roster count: 3)
  Player: Chico
  Player: Groucho
  Player: Harpo
Team name: Burns and Allen (roster count: 2)
  Player: George
  Player: Gracie
```

This example hints at what is possible in a commercial-grade, SOAP-based web service. Programmer-defined types such as Player and Team, along with arbitrary collections of these, can be arguments passed to or values returned from a web service so long as certain guidelines are followed. One guideline comes into play in this example. For the Team and the Player classes, the JavaBean properties are of types String or int; and a List, like any Java Collection, has a toArray method. In the end, a List<Team> reduces to arrays of simple types; in this case String instances or int values. The next chapter covers the details, in particular how the *wsgen* and the *wsimport* utilities facilitate the development of JWS services and clients.

Multithreading the Endpoint Publisher

In the examples so far, the Endpoint publisher has been single-threaded and, therefore, capable of handling only one client request at a time: the published service completes the processing of one request before beginning the processing of another request. If the processing of the current request hangs, then no subsequent request can be processed unless and until the hung request is processed to completion.

In production mode, the Endpoint publisher would need to handle concurrent requests so that several pending requests could be processed at the same time. If the underlying computer system is, for example, a symmetric multiprocessor (SMP), then separate CPUs could process different requests concurrently. On a single-CPU machine, the concurrency would occur through time sharing; that is, each request would get a share of the available CPU cycles so that several requests would be in some stage of processing at any given time. In Java, concurrency is achieved through multithreading. At issue, then, is how to make the Endpoint publisher multithreaded. The JWS framework supports Endpoint multithreading without forcing the programmer to work with difficult, error-prone constructs such as the synchronized block or the wait and notify method invocations.

An Endpoint object has an Executor property defined with the standard get/set methods. An Executor is an object that executes Runnable tasks; for example, standard Java Thread instances. (The Runnable interface declares only one method whose declaration is public void run().) An Executor object is a nice alternative to Thread instances, as the Executor provides high-level constructs for submitting and managing tasks that are to be executed concurrently. The first step to making the Endpoint publisher multithreaded is thus to create an Executor class such as the following very basic one:

```
package ch01.ts;

import java.util.concurrent.ThreadPoolExecutor;
import java.util.concurrent.LinkedBlockingQueue;
import java.util.concurrent.TimeUnit;
import java.util.concurrent.locks.ReentrantLock;
import java.util.concurrent.locks.Condition;

public class MyThreadPool extends ThreadPoolExecutor {
    private static final int pool_size = 10;
    private boolean is_paused;
    private ReentrantLock pause_lock = new ReentrantLock();
    private Condition unpaused = pause_lock.newCondition();

    public MyThreadPool(){
       super(pool_size,         // core pool size
             pool_size,         // maximum pool size
             0L,                // keep-alive time for idle thread
             TimeUnit.SECONDS, // time unit for keep-alive setting
             new LinkedBlockingQueue<Runnable>(pool_size)); // work queue
    }

    // some overrides
    protected void beforeExecute(Thread t, Runnable r) {
       super.beforeExecute(t, r);
       pause_lock.lock();
       try {
          while (is_paused) unpaused.await();
       }
```

```
            catch (InterruptedException e) { t.interrupt(); }
            finally { pause_lock.unlock(); }
        }

        public void pause() {
            pause_lock.lock();
            try {
                is_paused = true;
            }
            finally { pause_lock.unlock(); }
        }

        public void resume() {
            pause_lock.lock();
            try {
                is_paused = false;
                unpaused.signalAll();
            }
            finally { pause_lock.unlock(); }
        }
    }
```

The class MyThreadPool creates a pool of 10 threads, using a fixed-size queue to store the threads that are created under the hood. If the pooled threads are all in use, then the next task in line must wait until one of the busy threads becomes available. All of these management details are handled automatically. The MyThreadPool class overrides a few of the available methods to give the flavor.

A MyThreadPool object can be used to make a multithreaded Endpoint publisher. Here is the revised publisher, which now consists of several methods to divide the work:

```
package ch01.ts;

import javax.xml.ws.Endpoint;

class TimePublisherMT { // MT for multithreaded
    private Endpoint endpoint;

    public static void main(String[ ] args) {
        TimePublisherMT self = new TimePublisherMT();
        self.create_endpoint();
        self.configure_endpoint();
        self.publish();
    }
    private void create_endpoint() {
        endpoint = Endpoint.create(new TimeServerImpl());
    }
    private void configure_endpoint() {
        endpoint.setExecutor(new MyThreadPool());
    }
```

```
    private void publish() {
        int port = 8888;
        String url = "http://localhost:" + port + "/ts";
        endpoint.publish(url);
        System.out.println("Publishing TimeServer on port " + port);
    }
}
```

Once the `ThreadPoolExecutor` has been coded, all that remains is to set the `Endpoint` publisher's executor property to an instance of the worker class. The details of thread management do not intrude at all into the publisher.

The multithreaded `Endpoint` publisher is suited for lightweight production, but this publisher is not a service *container* in the true sense; that is, a software application that readily can deploy many web services at the same port. A web container such as Tomcat, which is the reference implementation, is better suited to publish multiple web services. Tomcat is introduced in later examples.

What's Next?

A SOAP-based web service should provide, as a WSDL document, a service contract for its potential clients. So far we have seen how a Perl, a Ruby, and a Java client can request the WSDL at runtime for use in the underlying SOAP libraries. Chapter 2 studies the WSDL more closely and illustrates how it may be used to generate client-side artifacts such as Java classes, which in turn ease the coding of web service clients. The Java clients in Chapter 2 will not be written from scratch, as is our first Java client. Instead such clients will be written with the considerable aid of the *wsimport* utility, as was the `TeamClient` shown earlier. Chapter 2 also introduces *JAX-B* (Java API for XML-Binding), a collection of Java packages that coordinate Java data types and XML data types. The *wsgen* utility generates JAX-B artifacts that play a key role in this coordination; hence, *wsgen* also will get a closer look.

All About WSDLs

What Good Is a WSDL?

The usefulness of WSDLs, the service contracts for SOAP-based web services, is shown best through examples. The original Java client against the `TimeServer` service invokes the `Service.create` method with two arguments: a URL, which provides the endpoint at which the service can be accessed, and an XML-qualified name (a Java `QName`), which in turn consists of the service's local name (in this case, `TimeServerImplService`) and a namespace identifier (in this case, the URI `http://ts.ch01/`). Here is the relevant code without the comments:

```
URL url = new URL("http://localhost:9876/ts?wsdl");
QName qname = new QName("http://ts.ch01/", "TimeServerImplService");
Service service = Service.create(url, qname);
```

Note that the automatically generated namespace URI *inverts* the package name of the service implementation bean (SIB), `ch01.ts.TimeServerImpl`. The package `ch01.ts` becomes `ts.ch01` in the URI. This detail is critical. If the first argument to the `QName` constructor is changed to the URI `http://ch01.ts/`, the Java `TimeClient` throws a exception, complaining that the service's automatically generated WSDL does not describe a service with this namespace URI. The programmer must figure out the namespace URI—presumably by inspecting the WSDL! By the way, even the trailing slash in the URI is critical. The URI `http://ts.ch01`, with a slash missing at the end, causes the same exception as does `http://ch01.ts/`, with *ch01* and *ts* in the wrong order.

The same point can be illustrated with a revised Perl client, which accesses the web service but without requesting its WSDL, as shown in Example 2-1.

Example 2-1. A revised Perl client for the TimeServer service

```
#!/usr/bin/perl -w

use SOAP::Lite;

my $endpoint = 'http://127.0.0.1:9876/ts'; # endpoint
my $uri      = 'http://ts.ch01/';          # namespace
```

```
my $client = SOAP::Lite->uri($uri)->proxy($endpoint);

my $response = $client->getTimeAsString()->result();
print $response, "\n";
$response = $client->getTimeAsElapsed()->result();
print $response, "\n";
```

The revised Perl client is functionally equivalent to the original. However, the revised Perl client must specify the service's namespace URI: http://ts.ch01/. No other URI would work because the Java-generated WSDL produces exactly this one. In effect, the web service is accessible through a pair: a service endpoint (URL) and a service name-space (URI). It is harder to code the Perl client in Example 2-1 than the original Perl client. The revised client requires the programmer to know the service namespace URI in addition to the service endpoint URL. The original Perl client sidesteps the problem by requesting the WSDL document, which includes the service URI. The original Perl client, which gets the URI from the consumed WSDL document, is the easier way to go.

Generating Client-Support Code from a WSDL

Java has a *wsimport* utility that eases the task of writing a Java client against a SOAP-based web service. The utility generates client-support code or *artifacts* from the service contract, the WSDL document. At a command prompt, the command:

```
% wsimport
```

displays a short report on how the utility can be used. The first example with *wsimport* produces client artifacts against the TimeServer service.

After the ch01.ts.TimeServerPublisher application has been started, the command:

```
% wsimport -keep -p client http://localhost:9876/ts?wsdl
```

generates two source and two compiled files in the subdirectory *client*. The URL at the end—the same one used in the original Perl, Ruby, and Java clients to request the service's WSDL document—gives the location of the service contract. The option -p specifies the Java package in which the generated files are to be placed, in this case a package named client; the package name is arbitrary and the utility uses or creates a subdirectory with the package name. The option -keep indicates that the source files should be kept, in this case for inspection. The -p option is important because *wsimport* generates the file *TimeServer.class*, which has the same name as the compiled versions of the original service endpoint interface (SEI). If a package is not specified with the *wsimport* utility, then the default package is the package of the service implementation, in this case ch01.ts. In short, using the -p option prevents the compiled SEI file from being overwritten by the file generated with the *wsimport* utility. If a local copy of the WSDL document is available (for instance, the file named *ts.wsdl*), then the command would be:

```
% wsimport -keep -p client ts.wsdl
```

Examples 2-2 and 2-3 are the two source files, with comments removed, that the *wsimport* command generates.

Example 2-2. The wsimport-generated TimeServer

```java
package client;

import javax.jws.WebMethod;
import javax.jws.WebResult;
import javax.jws.WebService;
import javax.jws.soap.SOAPBinding;

@WebService(name = "TimeServer", targetNamespace = "http://ts.ch01/")
@SOAPBinding(style = SOAPBinding.Style.RPC)
public interface TimeServer {
    @WebMethod
    @WebResult(partName = "return")
    public String getTimeAsString();

    @WebMethod
    @WebResult(partName = "return")
    public long getTimeAsElapsed();
}
```

Example 2-3. The wsimport-generated TimeServerImplService

```java
package client;

import java.net.MalformedURLException;
import java.net.URL;
import javax.xml.namespace.QName;
import javax.xml.ws.Service;
import javax.xml.ws.WebEndpoint;
import javax.xml.ws.WebServiceClient;

@WebServiceClient(name = "TimeServerImplService",
                  targetNamespace = "http://ts.ch01/",
                  wsdlLocation = "http://localhost:9876/ts?wsdl")
public class TimeServerImplService extends Service {
    private final static URL TIMESERVERIMPLSERVICE_WSDL_LOCATION;

    static {
        URL url = null;
        try {
            url = new URL("http://localhost:9876/ts?wsdl");
        }
        catch (MalformedURLException e) {
            e.printStackTrace();
        }
        TIMESERVERIMPLSERVICE_WSDL_LOCATION = url;
    }

    public TimeServerImplService(URL wsdlLocation, QName serviceName) {
        super(wsdlLocation, serviceName);
    }
```

```
    public TimeServerImplService() {
        super(TIMESERVERIMPLSERVICE_WSDL_LOCATION,
            new QName("http://ts.ch01/", "TimeServerImplService"));
    }

    @WebEndpoint(name = "TimeServerImplPort")
    public TimeServer getTimeServerImplPort() {
        return (TimeServer)super.getPort(new QName("http://ts.ch01/",
                                    "TimeServerImplPort"),
                            TimeServer.class);
    }
}
```

Three points about these generated source files deserve mention. First, the interface
client.TimeServer declares the very same methods as the original SEI, TimeServer. The
methods are the web service operation getTimeAsString and the operation
getTimeAsElapsed. Second, the class client.TimeServerImplService has a no-argument
constructor that constructs the very same Service object as the original Java client
TimeClient. Third, TimeServerImplService encapsulates the getTimeServerImplPort
method, which returns an instance of type client.TimeServer, which in turn supports
invocations of the two web service operations. Together the two generated types, the
interface client.TimeServer and the class client.TimeServerImplService, ease the task
of writing a Java client against the web service. Example 2-4 shows a client that uses
the client-support code from the *wsimport* utility.

Example 2-4. A Java client that uses wsimport artifacts

```
package client;

class TimeClientWSDL {
    public static void main(String[ ] args) {
        // The TimeServerImplService class is the Java type bound to
        // the service section of the WSDL document.
        TimeServerImplService service = new TimeServerImplService();

        // The TimeServer interface is the Java type bound to
        // the portType section of the WSDL document.
        TimeServer eif = service.getTimeServerImplPort();

        // Invoke the methods.
        System.out.println(eif.getTimeAsString());
        System.out.println(eif.getTimeAsElapsed());
    }
}
```

The client in Example 2-4 is functionally equivalent to the original TimeClient, but this
client is far easier to write. In particular, troublesome yet critical details such as the
appropriate QName and service endpoint now are hidden in the *wsimport*-generated class,
client.TempServerImplService. The idiom that is illustrated here works in general for

writing clients with help from WSDL-based artifacts such as `TimeServer` and
`TimeServerImplService`:

- First, construct a `Service` object using one of two constructors in the *wsimport*-generated class, in this example `client.TimeServerImplService`. The no-argument constructor is preferable because of its simplicity. However, a two-argument constructor is also available in case the web service's namespace (URI) or the service endpoint (URL) have changed. Even in this case, however, it would be advisable to regenerate the WSDL-based Java files with another use of the *wsimport* utility.

- Invoke the `get...Port` method on the constructed `Service` object, in this example, the method `getTimeServerImplPort`. The method returns an object that encapsulates the web service operations, in this case `getTimeAsString` and `getTimeAsElapsed`, declared in the original SEI.

The @WebResult Annotation

The WSDL-based `client.TimeServer` interface introduces the `@WebResult` annotation. To show how this annotation works, Example 2-5 is a revised version of the `ch01.ts.TimeServer` SEI (with comments removed).

Example 2-5. A more annotated version of the TimeServer service

```
package ch01.ts;  // time server

import javax.jws.WebService;
import javax.jws.WebMethod;
import javax.jws.WebResult;
import javax.jws.soap.SOAPBinding;
import javax.jws.soap.SOAPBinding.Style;

@WebService
@SOAPBinding(style = Style.RPC) // more on this later
public interface TimeServer {
    @WebMethod
    @WebResult(partName = "time_response")
    String getTimeAsString();

    @WebMethod
    @WebResult(partName = "time_response")
    long getTimeAsElapsed();
}
```

The `@WebResult` annotates the two web service operations. In the resulting WSDL, the
`message` section reflects this change:

```
<message name="getTimeAsString"></message>
<message name="getTimeAsStringResponse">
    <part name="time_response" type="xsd:string"></part>
</message>
<message name="getTimeAsElapsed"></message>
```

```
<message name="getTimeAsElapsedResponse">
    <part name="time_response" type="xsd:long"></part>
</message>
```

Note that `time_response` now replaces `return` from the original WDSL. The SOAP response document from the web service likewise reflects the change:

```
<?xml version="1.0" ?>
<soapenv:Envelope
    xmlns:soapenv="http://schemas.xmlsoap.org/soap/envelope/"
    xmlns:xsd="http://www.w3.org/2001/XMLSchema">
  <soapenv:Body>
    <ans:getTimeAsStringResponse xmlns:ans="http://ts.ch01/">
      <time_response>
          Thu Mar 27 21:20:09 CDT 2008
      </time_response>
    </ans:getTimeAsStringResponse>
  </soapenv:Body>
</soapenv:Envelope>
```

Once again, the `time_response` tag replaces the `return` tag from the original SOAP response. If the `@WebResult` annotation were applied, say, only to the `getTimeAsString` operation, then the SOAP response for this operation would use the `time_response` tag but the response for the `getTimeAsElapsed` operation still would use the default `return` tag.

The point of interest is that various annotations are available to determine what the generated WSDL document will look like. Such annotations will be introduced in small doses. It is easiest to minimize the use of annotations, using only the ones that serve some critical purpose. The `TimeServer` works exactly the same whether the `time_response` or the default `return` tag is used in the WSDL and in the SOAP response; hence, there is usually no need for programmer-generated code to use the `@WebResult` annotation.

Various commercial-grade WSDLs are available for generating Java support code. An example that will be introduced shortly uses a WSDL from Amazon's Associates Web Service, more popularly known as Amazon's E-Commerce service. However, the Amazon example first requires a closer look at WSDL structure. My plan is to keep the tedious details to a minimum. The goal of the next section is to move from basic WSDL structure to the unofficial but popular distinction between *wrapped* and *unwrapped* SOAP message bodies.

WSDL Structure

At a high level, a WSDL document is a contract between a service and its consumers. The contract provides such critical information as the service endpoint, the service operations, and the data types required for these operations. The service contract also indicates, in describing the messages exchanged in the service, the underlying service pattern, for instance, request/response or solicit/response. The outermost element (called the *document* or *root* element) in a WSDL is named `definitions` because the WSDL provides definitions grouped into the following sections:

- The `types` section, which is optional, provides data type definitions under some data type system such as XML Schema. A particular document that defines data types is an *XSD* (XML Schema Definition). The `types` section holds, points to, or imports an XSD. If the `types` section is empty, as in the case of the `TimeServer` service, then the service uses only simple data types such as `xsd:string` and `xsd:long`.

  Although the WSDL 2.0 specification allows for alternatives to XML Schema (see *http://www.w3.org/TR/wsdl20-altschemalangs*), XML Schema is the default and the dominant type system used in WSDLs. Accordingly, the following examples assume XML Schema unless otherwise noted.

- The `message` section defines the messages that implement the service. Messages are constructed from data types either defined in the immediately preceding section or, if the `types` section is empty, available as defaults. Further, the order of the messages indicates the service pattern. Recall that, for messages, the directional properties `in` and `out` are from the service's perspective: an `in` message is to the service, whereas an `out` message is from the service. Accordingly, the message order `in/out` indicates the request/response pattern, whereas the message order `out/in` indicates the solicit/response pattern. For the `TimeServer` service, there are four messages: a request and a response for the two operations, `getTimeAsString` and `getTimeAsElapsed`. The `in/out` order in each pair indicates a request/response pattern for the web service operations.

- The `portType` section presents the service as named operations, with each operation as one or more messages. Note that the operations are named after methods annotated as `@WebMethods`, a point to be discussed in detail shortly. A web service's `portType` is akin to a Java interface in presenting the service abstractly, that is, with no implementation details.

- The `binding` section is where the WSDL definitions go from the abstract to the concrete. A WSDL binding is akin to a Java implementation of an interface (that is, a WSDL `portType`). Like a Java implementation class, a WSDL `binding` provides important concrete details about the service. The `binding` section is the most complicated one because it must specify these implementation details of a service defined abstractly in the `portType` section:

  — The transport protocol to be used in sending and receiving the underlying SOAP messages. Either HTTP or *SMTP* (Simple Mail Transport Protocol) may be used for what is called the *application-layer* protocol; that is, the protocol for transporting the SOAP messages that implement the service. HTTP is by far the more popular choice. The WSDL for the `TimeServer` service contains this segment:

    ```
    <soap:binding style="rpc"
              transport="http://schemas.xmlsoap.org/soap/http">
    ```

The value of the **transport** attribute signals that the service's SOAP messages will be sent and received over HTTP, which is captured in the slogan *SOAP over HTTP*.

— The **style** of the service, shown earlier as the value of the **style** attribute, takes either **rpc** or **document** as a value. The **document** style is the default, which explains why the SEI for the **TimeServer** service contains the annotation:

```
@SOAPBinding(style = Style.RPC)
```

This annotation forces the **style** attribute to have the value **rpc** in the Java-generated WSDL. The difference between the **rpc** and the **document** style will be clarified shortly.

— The data format to be used in the SOAP messages. There are two choices, **literal** and **encoded**. These choices also will be clarified shortly.

- The **service** section specifies one or more endpoints at which the service's functionality, the sum of its operations, is available. In technical terms, the **service** section lists one or more **port** elements, where a **port** consists of a **portType** (interface) together with a corresponding **binding** (implementation). The term **port** derives from distributed systems. An application hosted at a particular network address (for instance, 127.0.0.1) is available to clients, local or remote, through a specified port. For example, the **TimeServer** application is available to clients at port 9876.

The tricky part of the binding section involves the possible combinations of the style and the use attributes. The next subsection looks more closely at the relationships between these attributes.

A Closer Look at WSDL Bindings

In the WSDL **binding** section, the **style** attribute has **rpc** and **document** as possible values, with **document** as the default. The **use** attribute has **literal** and **encoded** as possible values, with **literal** as the default. In theory, there are four possibilities, as shown in Table 2-1.

Table 2-1. Possible combinations of style and use

style	use
document	literal
document	encoded
rpc	literal
rpc	encoded

Of the four possible combinations listed in Table 2-1, only two occur regularly in contemporary SOAP-based web services: **document/literal** and **rpc/literal**. For one thing, the `encoded` use, though valid in a WSDL document, does not comply with the *WS-I* (Web Services-Interoperability) guidelines (see *http://www.ws-i.org*). As the name indicates, the WS-I initiative is meant to help software architects and developers produce web services that can interoperate seamlessly despite differences in platforms and programming languages.

Before going any further into the details, it will be helpful to have a sample WSDL for a `document`-style service, which then can be contrasted with the WSDL for the `rpc`-style `TimeServer` service. The `use` attribute will be clarified after the `style` attribute.

The `TimeServer` service can be changed to a `document`-style service in two quick steps. The first is to comment out the line:

```
@SOAPBinding(style = Style.RPC)
```

in the SEI source, `ch02.ts.TimeServer.java`, before recompiling. (The package has been changed from `ch01.ts` to `ch02.ts` to reflect that this is Chapter 2.) Commenting out this annotation means that the default style, `Style.DOCUMENT`, is in effect. The second step is to execute, in the working directory, the command:

```
% wsgen -keep -cp . ch02.ts.TimeServerImpl
```

The revised SEI and the corresponding SIB must be recompiled before the *wsgen* utility is used so that the `document`-style version is current. The *wsgen* utility then generates four source and four compiled files in the subdirectory *ch02/ts/jaxws*. These files provide the data types needed, in the `document`-style service, to produce the WSDL automatically. For example, among the files are source and compiled versions of the classes `GetTimeAsElapsed`, which is a request type, and `GetTimeAsElapsedResponse`, which is a response type. As expected, these two types support requests to and responses from the service operation `getTimeAsElapsed`. The *wsgen* utility generates comparable types for the `getTimeAsString` operation as well. The *wsgen* utility will be examined again later in this chapter.

The revised program can be published as before with the `TimeServerPublisher`, and the WSDL then is available at the published URL *http://localhost:9876/ts?wsdl*. The contrasting WSDLs now can be used to explain in detail how the `document` and `rpc` styles differ. There is one other important change to the `TimeServer`, which is reflected in the WSDL: the package name changes from `ch01.ts` to `ch02.ts`, a change that is reflected in a namespace throughout the WSDL.

Key Features of Document-Style Services

The `document` style indicates that a SOAP-based web service's underlying messages contain full XML documents; for instance, a company's product list as an XML document or a customer's order for some products from this list as another XML document. By contrast, the `rpc` style indicates that the underlying SOAP messages contain parameters in the request messages and return values in the response messages. Following is a segment of the `rpc`-style WSDL for the original `TimeServer` service:

```
<types></types>
<message name="getTimeAsString"></message>
<message name="getTimeAsStringResponse">
  <part name="time_response" type="xsd:string"></part>
</message>
<message name="getTimeAsElapsed"></message>
<message name="getTimeAsElapsedResponse">
  <part name="time_response" type="xsd:long"></part>
</message>
```

The types section is empty because the service uses, as return values, only the simple types xsd:string and xsd:long. The simple types do not require a definition in the WSDL's types section. Further, the message names show their relationships to the corresponding Java @WebMethods; in this case, the method getTimeAsString and the method getTimeAsElapsed. The response messages have a part subelement that gives the data type of the returned value, but because the request messages expect no arguments, the request messages do not need part subelements to describe parameters.

By contrast, here is the same WSDL segment for the document-style version of the TimeServer service:

```
<types>
  <xsd:schema>
    <xsd:import schemaLocation="http://localhost:9876/ts?xsd=1"
        namespace="http://ts.ch02/">
    </xsd:import>
  </xsd:schema>
</types>
<message name="getTimeAsString">
  <part element="tns:getTimeAsString" name="parameters"></part>
</message>
<message name="getTimeAsStringResponse">
  <part element="tns:getTimeAsStringResponse" name="parameters"></part>
</message>
<message name="getTimeAsElapsed">
  <part element="tns:getTimeAsElapsed" name="parameters"></part>
</message>
<message name="getTimeAsElapsedResponse">
  <part element="tns:getTimeAsElapsedResponse" name="parameters"></part>
</message>
```

The types now contains an import directive for the associated XSD document. The URL for the XSD is *http://localhost:9876/ts?xsd=1*. Example 2-6 is the XSD associated with the WSDL.

Example 2-6. The XSD for the TimeServer WSDL

```
<?xml version="1.0" encoding="UTF-8"?>
<xs:schema xmlns:tns="http://ts.ch02/"
           xmlns:xs="http://www.w3.org/2001/XMLSchema"
           targetNamespace="http://ts.ch02/" version="1.0">
  <xs:element name="getTimeAsElapsed"
              type="tns:getTimeAsElapsed">
  </xs:element>
```

```
<xs:element name="getTimeAsElapsedResponse"
            type="tns:getTimeAsElapsedResponse">
</xs:element>
<xs:element name="getTimeAsString"
            type="tns:getTimeAsString">
</xs:element>
<xs:element name="getTimeAsStringResponse"
            type="tns:getTimeAsStringResponse">
</xs:element>
<xs:complexType name="getTimeAsString"></xs:complexType>
<xs:complexType name="getTimeAsStringResponse">
  <xs:sequence>
    <xs:element name="return"
                type="xs:string" minOccurs="0">
    </xs:element>
  </xs:sequence>
</xs:complexType>
<xs:complexType name="getTimeAsElapsed"></xs:complexType>
<xs:complexType name="getTimeAsElapsedResponse">
  <xs:sequence>
    <xs:element name="return" type="xs:long"></xs:element>
  </xs:sequence>
</xs:complexType>
</xs:schema>
```

The XSD document defines four complex types whose names (for instance, the class `getTimeAsElapsedResponse`) are the names of corresponding messages in the `message` section. Under the `rpc`-style, the messages carry the names of the `@WebMethods`; under the `document`-style, there is an added level of complexity in that the messages carry the names of XSD types defined in the WSDL's **types** section.

The request/response pattern in a web service is possible under either the `document` or the `rpc` style, although the style named *rpc* obviously underscores this pattern. Under the `rpc` style, the messages are named but not explicitly typed; under the `document` style, the messages are explicitly typed in an XSD document.

The `document` style deserves to be the default. This style can support services with rich, explicitly defined data types because the service's WSDL can define the required types in an XSD document. Further, the `document` style can support any service pattern, including request/response. Indeed, the SOAP messages exchanged in the `TimeServer` application look the same regardless of whether the `style` is `rpc` or `document`. From an architectural perspective, the `document` style is the simpler of the two in that the body of a SOAP message is a self-contained, precisely defined document. The `rpc` style requires messages with the names of the associated operations (in Java, the `@WebMethods`) with parameters as subelements. From a developer perspective, however, the `rpc` style is the simpler of the two because the *wsgen* utility is not needed to generate Java types that correspond to XML Schema types. (The current Metro release generates the *wsgen* artifacts automatically. The details follow shortly.)

Finally, the `use` attribute determines how the service's data types are to be encoded and decoded. The WSDL service contract has to specify how data types used in the implementation language (for instance, Java) are serialized to WSDL-compliant types (by default, XML Schema types). On the client side, the WSDL-compliant types then must be deserialized into client-language types (for instance, C or Ruby types). The setting:

```
use = 'literal'
```

means that the service's type definitions *literally* follow the WSDL document's XML Schema. By contrast, the setting:

```
use = 'encoded'
```

means that the service's type definitions come from encoding rules, typically the encoding rules in the SOAP 1.1 specification. As noted earlier, the document/encoded and rpc/encoded combinations are not WS-I compliant. The real choices are, therefore, document/literal and rpc/literal.

Validating a SOAP Message Against a WSDL's XML Schema

One more point about rpc-style versus document-style bindings needs to be made. On the receiving end, a document-style SOAP message can be validated straightforwardly against the associated XML Schema. In rpc-style, by contrast, the validation is trickier precisely because there is no associated XML Schema. Example 2-7 is a short Java program that validates an arbitrary XML document against an arbitrary XSD document.

Example 2-7. A short program to validate an XML document

```java
import javax.xml.transform.stream.StreamSource;
import javax.xml.validation.SchemaFactory;
import javax.xml.validation.Schema;
import javax.xml.XMLConstants;
import javax.xml.validation.Validator;

class ValidateXML {
    public static void main(String[ ] args) {
        if (args.length != 2) {
            String msg = "\nUsage: java ValidateXML XMLfile XSDfile";
            System.err.println(msg);
            return;
        }
        try {
            // Read and validate the XML Schema (XSD document).
            final String schema_uri = XMLConstants.W3C_XML_SCHEMA_NS_URI;
            SchemaFactory factory = SchemaFactory.newInstance(schema_uri);
            Schema schema = factory.newSchema(new StreamSource(args[1]));
            // Validate the XML document against the XML Schema.
            Validator val = schema.newValidator();
            val.validate(new StreamSource(args[0]));
        }
        // Return on any validation error.
        catch(Exception e) {
            System.err.println(e);
```

```
        return;
    }
    System.out.println(args[0] + " validated against " + args[1]);
  }
}
```

Here is the extracted body of a SOAP response to the document-style TimeServer service:

```
<ns1:getTimeAsElapsedResponse xmlns:ns1="http://ts.ch02/">
   <return>1208229395922</return>
</ns1:getTimeAsElapsedResponse>
```

The body has been edited slightly by the addition of the namespace declaration xmlns:ns1="http://ts.ch02/, which sets ns1 as an alias or proxy for the namespace URI http://ts.ch02/. This URI does occur in the SOAP message but in the element SOAP::Envelope rather than in the ns1:getTimeAsElapsedResponse subelement. For a review of the full XSD, see Example 2-6. To keep the processing simple, the response body and the XSD are in the local files *body.xml* and *ts.xsd*, respectively. The command:

```
% java ValidateXML body.xml ts.xsd
```

validates the response's body against the XSD. If the return element in the body were changed to, say, foo bar, the attempted validation would produce the error message:

```
'foo bar' is not a valid value for 'integer'
```

thereby indicating that the SOAP message did not conform to the associated WSDL's XSD document.

The Wrapped and Unwrapped Document Styles

The document style is the default under the WS-I Basic profile for web services interoperability (see *http://www.ws-i.org/Profiles/BasicProfile-1.1.html*). This default style provides, through the XSD in the WSDL's types section, an explicit and precise definition of the data types in the service's underlying SOAP messages. The document style thereby promotes web service interoperability because a service client can determine precisely which data types are involved in the service and how the document contained in the body of an underlying SOAP message should be structured. However, the rpc style still has appeal in that the web service's operations have names linked directly to the underlying implementations; for example, to Java @WebMethods. For instance, the TimeServer service in the rpc style has a @WebMethod with the name getTimeAsString whose WSDL counterparts are the request message named getTimeAsString and the response message named getTimeAsStringResponse. The rpc style is programmer-friendly.

The gist of the wrapped convention, which is unofficial but widely followed, is to give a document-style service the look and feel of an rpc-style service. The wrapped convention seeks to combine the benefits of document and rpc styles.

To begin, Example 2-8 is an example of an unwrapped SOAP envelope, and Example 2-9 is an example of a wrapped SOAP envelope.

Example 2-8. Unwrapped document style

```
<?xml version="1.0" ?>
<!-- Unwrapped document style -->
<soapenv:Envelope
    xmlns:soapenv="http://schemas.xmlsoap.org/soap/envelope/"
    xmlns:xsd="http://www.w3.org/2001/XMLSchema">
  <soapenv:Body>
    <num1 xmlns:ans="http://ts.ch01/">27</num1>
    <num2 xmlns:ans="http://ts.ch01/">94</num2>
  </soapenv:Body>
</soapenv:Envelope>
```

Example 2-9. Wrapped document style

```
<?xml version="1.0" ?>
<!-- Wrapped document style -->
<soapenv:Envelope
    xmlns:soapenv="http://schemas.xmlsoap.org/soap/envelope/"
    xmlns:xsd="http://www.w3.org/2001/XMLSchema">
  <soapenv:Body>
    <addNums xmlns:ans="http://ts.ch01/">
      <num1>27</num1>
      <num2>94</num2>
    </addNums>
  </soapenv:Body>
</soapenv:Envelope>
```

The body of the unwrapped SOAP request envelope has two elements, named num1 and num2. These are numbers to be added. The SOAP body does *not* contain the name of the service operation that is to perform the addition and send the sum as a response. By contrast, the body of a wrapped SOAP request envelope has a single element named addNums, the name of the requested service operation, and two subelements, each holding a number to be added. The wrapped version makes the service operation explicit. The arguments for the operation are nested in an intuitive manner; that is, as subelements within the operation element addNums.

Guidelines for the wrapped document convention are straightforward. Here is a summary of the guidelines:

- The SOAP envelope's body should have only one part, that is, it should contain a single XML element with however many XML subelements are required. For example, even if a service operation expects arguments and returns a value, the parameters and return value do not occur as standalone XML elements in the SOAP body but, rather, as XML subelements within the main element. Example 2-9 illustrates with the addNums element as the single XML element in the SOAP body and the pair num1 and num2 as XML subelements.

- The relationship between the WSDL's XSD and the single XML element in the SOAP body is well defined. The `document`-style version of the `TimeServer` can be used to illustrate. In the XSD, there are four XSD `complexType` elements, each of which defines a data type. For example, there is a `complexType` with the name `getTimeAsString` and another with the name `getTimeAsStringResponse`. These definitions occur in roughly the bottom half of the XSD. In the top half are XML `element` definitions, each of which has a name and a type attribute. The `complexTypes` also have names, which are coordinated with the `element` names. For example, the `complexType` named `getTimeAsString` is matched with an element of the same name. Here is a segment of the XSD that shows the name coordination:

```
<xs:element name="getTimeAsString"
            type="tns:getTimeAsString">
</xs:element>
<xs:element name="getTimeAsStringResponse"
            type="tns:getTimeAsStringResponse">
</xs:element>
...
<xs:complexType name="getTimeAsString"></xs:complexType>
<xs:complexType name="getTimeAsStringResponse">
    <xs:sequence>
      <xs:element name="return"
                  type="xs:string" minOccurs="0">
      </xs:element>
    </xs:sequence>
</xs:complexType>
```

Further, each `complexType` is either empty (for instance, `getTimeAsString`) or contains an `xs:sequence` (for instance, `getTimeAsStringResponse`, which has an `xs:sequence` of one XML element). The `xs:sequence` contains typed arguments and typed returned values. The `TimeServer` example is quite simple in that the requests contain no arguments and the responses contain just one return value. Nonetheless, this segment of XSD shows the structure for the general case. For instance, if the `getTimeAsStringResponse` had several return values, then each would occur as an XML subelement within the `xs:sequence`. Finally, note that every XML element in this XSD segment (and, indeed, in the full XSD) is named and typed.

- The XML elements in the XSD serve as the *wrappers* for the SOAP message body. For the `ch01.ts.TimeServer`, a sample wrapped `document` is:

```
<?xml version="1.0" ?>
<soapenv:Envelope
    xmlns:soapenv="http://schemas.xmlsoap.or g/soap/envelope/"
    xmlns:xsd="http://www.w3.org/2001/XMLSchema">
  <soapenv:Body>
    <ans:getTimeAsElapsedResponse xmlns:ans="http://ts.ch01/">
      <return>1205030105192</return>
    </ans:getTimeAsElapsedResponse>
  </soapenv:Body>
</soapenv:Envelope>
```

This is the same kind of SOAP body generated for an rpc-style service, which is precisely the point. The difference, again, is that the wrapped document-style service, unlike its rpc-style counterpart, includes explicit type and format information in an XSD from the WSDL's types section.

- The request wrapper has the same name as the service operation (for instance, addNums in Example 2-9), and the response wrapper should be the request wrapper's name with Response appended (for instance, addNumsResponse).

- The WSDL portType section now has named operations (e.g., getTimeAsString) whose messages are *typed*. For instance, the input (request) message getTimeAsString has the type tns:getTimeAsString, which is defined as one of the four complexTypes in the WSDL's XSD. For the document-style version of the service, here is the portType segment:

```
<portType name="TimeServer">
  <operation name="getTimeAsString">
    <input message="tns:getTimeAsString"></input>
    <output message="tns:getTimeAsStringResponse"></output>
  </operation>
  <operation name="getTimeAsElapsed">
    <input message="tns:getTimeAsElapsed"></input>
    <output message="tns:getTimeAsElapsedResponse"></output>
  </operation>
</portType>
```

The wrapped document convention, despite its unofficial status, has become prevalent. JWS supports the convention. By default, a Java SOAP-based web service is wrapped *doc/lit*, that is, wrapped document style with literal encoding.

This excursion into the details of WSDL binding section will be helpful in the next example, which uses the *wsimport* utility to generate client-support code against Amazon's E-Commerce web service. The contrast between wrapped and unwrapped is helpful in understanding the client-support code.

Amazon's E-Commerce Web Service

The section title has the popular and formerly official name for one of the web services that Amazon hosts. The official name is now *Amazon Associates Web Service*. The service in question is accessible as SOAP-based or REST-style. The service is free of charge, but it does require registration at *http://affiliate-program.amazon.com/gp/asso ciates/join*. For the examples in this section, an Amazon *access key* (as opposed to the *secret access key* used to generate an authentication token) is required.

Amazon's E-Commerce service replicates the interactive experience at the Amazon website (*http://www.amazon.com*). For example, the service supports searching for items, bidding on items and putting items up for bid, creating a shopping cart and filling it with items, and so on. The two sample clients illustrate item search.

This section examines two Java clients against the Amazon E-Commerce service. Each client is generated with Java support code from the *wsimport* utility introduced earlier. The difference between the two clients refines the distinction between the wrapped and unwrapped conventions.

An E-Commerce Client in Wrapped Style

The Java support code for the client can be generated with the command:

```
% wsimport -keep -p awsClient \
http://ecs.amazonaws.com/AWSECommerceService/AWSECommerceService.wsdl
```

Recall that the `-p awsClient` part of the command generates a package (and, therefore, a subdirectory) named *awsClient*.

The source code for the first Amazon client, `AmazonClientW`, resides in the working directory; that is, the parent directory of *awsClient*. Example 2-10 is the application code, which searches for books about quantum gravity.

Example 2-10. An E-Commerce Java client in wrapped style

```java
import awsClient.AWSECommerceService;
import awsClient.AWSECommerceServicePortType;
import awsClient.ItemSearchRequest;
import awsClient.ItemSearch;
import awsClient.Items;
import awsClient.Item;
import awsClient.OperationRequest;
import awsClient.SearchResultsMap;
import javax.xml.ws.Holder;
import java.util.List;
import java.util.ArrayList;

class AmazonClientW { // W is for Wrapped style
    public static void main(String[ ] args) {
      if (args.length < 1) {
        System.err.println("Usage: java AmazonClientW <access key>");
        return;
      }
      final String access_key = args[0];

      // Construct a service object to get the port object.
      AWSECommerceService service = new AWSECommerceService();
      AWSECommerceServicePortType port = service.getAWSECommerceServicePort();

      // Construct an empty request object and then add details.
      ItemSearchRequest request = new ItemSearchRequest();
      request.setSearchIndex("Books");
      request.setKeywords("quantum gravity");

      ItemSearch search = new ItemSearch();
      search.getRequest().add(request);
      search.setAWSAccessKeyId(access_key);
```

```
Holder<OperationRequest> operation_request = null;
Holder<List<Items>> items = new Holder<List<Items>>();

port.itemSearch(search.getMarketplaceDomain(),
                search.getAWSAccessKeyId(),
                search.getSubscriptionId(),
                search.getAssociateTag(),
                search.getXMLEscaping(),
                search.getValidate(),
                search.getShared(),
                search.getRequest(),
                operation_request,
                items);

        // Unpack the response to print the book titles.
        Items retval = items.value.get(0);        // first and only Items element
        List<Item> item_list = retval.getItem(); // list of Item subelements
        for (Item item : item_list)               // each Item in the list
          System.out.println(item.getItemAttributes().getTitle());
    }
}
```

The code is compiled and executed in the usual way but requires your access key (a string such as 1A67QRNF7AGRQ1XXMJ07) as a command-line argument. On a sample run, the output for an item search among books on the string quantum gravity was:

```
The Trouble With Physics
The Final Theory: Rethinking Our Scientific Legacy
Three Roads to Quantum Gravity
Keeping It Real (Quantum Gravity, Book 1)
Selling Out (Quantum Gravity, Book 2)
Mr Tompkins in Paperback
Head First Physics
Introduction to Quantum Effects in Gravity
The Large, the Small and the Human Mind
Feynman Lectures on Gravitation
```

The AmazonClientW client is not intuitive. Indeed, the code uses relatively obscure types such as Holder. The itemSearch method, which does the actual search, takes 10 arguments, the last of which, named items in this example, holds the service response. This response needs to be unpacked in order to get a list of the book titles returned from the search. The next client is far simpler. Before looking at the simpler client, however, it will be instructive to consider what makes the first client so tricky.

In the binding section of the WSDL for the E-Commerce service, the style is set to the default, document, and the encoding is likewise set to the default, literal. Further, the wrapped convention is in play, as this segment from the XSD illustrates:

```
<xs:element name="ItemSearch">
  <xs:complexType>
    <xs:sequence>
      <xs:element name="MarketplaceDomain"
                  type="xs:string" minOccurs="0"/>
```

```
            <xs:element name="AWSAccessKeyId"
                        type="xs:string" minOccurs="0"/>
            <xs:element name="SubscriptionId"
                        type="xs:string" minOccurs="0"/>
            <xs:element name="AssociateTag"
                        type="xs:string" minOccurs="0"/>
            <xs:element name="XMLEscaping"
                        type="xs:string" minOccurs="0"/>
            <xs:element name="Validate"
                        type="xs:string" minOccurs="0"/>
            <xs:element name="Shared"
                        type="tns:ItemSearchRequest" minOccurs="0"/>
            <xs:element name="Request"
                        type="tns:ItemSearchRequest" minOccurs="0" maxOccurs="unbounded"/>
        </xs:sequence>
      </xs:complexType>
    </xs:element>
```

This XSD segment defines the ItemSearch element, which is the wrapper type in the body of a SOAP request. Here is a segment from the WSDL's message section, which shows that the request message is the type defined above:

```
<message name="ItemSearchRequestMsg">
    <part name="body" element="tns:ItemSearch"/>
</message>
```

For the service response to an ItemSearch, the wrapper type defined in the XSD is:

```
<xs:element name="ItemSearchResponse">
  <xs:complexType>
    <xs:sequence>
      <xs:element ref="tns:OperationRequest" minOccurs="0"/>
      <xs:element ref="tns:Items" minOccurs="0" maxOccurs="unbounded"/>
    </xs:sequence>
  </xs:complexType>
</xs:element>
```

The impact of these WSDL definitions is evident in the SOAP request message from a client and the SOAP response message from the server. Example 2-11 shows a SOAP request from a sample run.

Example 2-11. A SOAP request against Amazon's E-Commerce service

```
<?xml version="1.0" ?>
<soapenv:Envelope
  xmlns:soapenv="http://schemas.xmlsoap.org/soap/envelope/"
  xmlns:xsd="http://www.w3.org/2001/XMLSchema"
  xmlns:ns1="http://webservices.amazon.com/AWSECommerceService/2008-03-03">
  <soapenv:Body>
    <ns1:ItemSearch>
      <ns1:AWSAccessKeyId>...</ns1:AWSAccessKeyId>
      <ns1:Request>
          <ns1:Keywords>quantum gravity</ns1:Keywords>
          <ns1:SearchIndex>Books</ns1:SearchIndex>
      </ns1:Request>
    </ns1:ItemSearch>
```

```
        </soapenv:Body>
</soapenv:Envelope>
```

The `ItemSearch` wrapper is the single XML element in the SOAP body and this element has two subelements, one with the name `ns1:AWSAccessKeyId` and the other with the name `ns1:Request`, whose subelements specify the search string (in this case, quantum gravity) and the search category (in this case, `Books`).

Here is part of the SOAP response for the request above:

```
<?xml version="1.0" encoding="UTF-8"?>
<SOAP-ENV:Envelope
    xmlns:SOAP-ENV="http://schemas.xmlsoap.org/soap/envelope/"
    xmlns:SOAP-ENC="http://schemas.xmlsoap.org/soap/encoding/"
    xmlns:xsi="http://www.w3.org/2001/XMLSchema-instance"
    xmlns:xsd="http://www.w3.org/2001/XMLSchema">
  <SOAP-ENV:Body>
    <ItemSearchResponse
        xmlns="http://webservices.amazon.com/AWSECommerceService/2008-03-03">
      <OperationRequest>
        <HTTPHeaders>
          <Header Name="UserAgent" Value="Java/1.6.0"></Header>
        </HTTPHeaders>
        <RequestId>004ON1YEKVOCRCT2B5PR</RequestId>
        <Arguments>
          <Argument Name="Service" Value="AWSECommerceService"></Argument>
        </Arguments>
        <RequestProcessingTime>0.0566580295562744</RequestProcessingTime>
      </OperationRequest>
      <Items>
        <Request>
          <IsValid>True</IsValid>
          <ItemSearchRequest>
            <Keywords>quantum gravity</Keywords>
            <SearchIndex>Books</SearchIndex>
          </ItemSearchRequest>
        </Request>
        <TotalResults>207</TotalResults>
        <TotalPages>21</TotalPages>
        <Item>
          <ASIN>061891868X</ASIN>
          <DetailPageURL>http://www.amazon.com/gp/redirect.html...
          </DetailPageURL>
          <ItemAttributes>
            <Author>Lee Smolin</Author>
            <Manufacturer>Mariner Books</Manufacturer>
            <ProductGroup>Book</ProductGroup>
            <Title>The Trouble With Physics</Title>
          </ItemAttributes>
        </Item>
        ...
      </Items>
    </ItemSearchResponse>
  </SOAP-ENV:Body>
</SOAP-ENV:Envelope>
```

The SOAP body now consists of a single element named `ItemSearchResponse`, which is defined as a type in the service's WSDL. This element contains various subelements, the most interesting of which is named `Items`. The `Items` subelement contains multiple `Item` subelements, one apiece for a book on quantum gravity. Only one `Item` is shown, but this is enough to see the structure of the response document. The code near the end of the `AmazonClientW` reflects the structure of the SOAP response: first the `Items` element is extracted, then a list of `Item` subelements, each of which contains a book's title as an XML attribute.

Now we can look at a *wsimport*-generated artifact for the E-Commerce service, in particular at the annotations on the `itemSearch` method with its 10 arguments. The method is declared in the `AWSECommerceServicePortType` interface, which declares a single `@WebMethod` for each of the E-Commerce service's operations. Example 2-12 shows the declaration of interest.

Example 2-12. Java code generated from Amazon's E-Commerce WSDL

```
@WebMethod(operationName = "ItemSearch",
           action = "http://soap.amazon.com")
@RequestWrapper(localName = "ItemSearch",
                targetNamespace =
                "http://webservices.amazon.com/AWSECommerceService/2008-03-03",
                className = "awsClient.ItemSearch")
@ResponseWrapper(localName = "ItemSearchResponse",
                 targetNamespace =
                 "http://webservices.amazon.com/AWSECommerceService/2008-03-03",
                 className = "awsClient.ItemSearchResponse")
public void itemSearch(
  @WebParam(name = "MarketplaceDomain",
            targetNamespace =
            "http://webservices.amazon.com/AWSECommerceService/2008-03-03")
  String marketplaceDomain,
  @WebParam(name = "AWSAccessKeyId",
            targetNamespace =
            "http://webservices.amazon.com/AWSECommerceService/2008-03-03")
  String awsAccessKeyId,
  ...
  ItemSearchRequest shared,
  @WebParam(name = "Request",
            targetNamespace =
            "http://webservices.amazon.com/AWSECommerceService/2008-03-03")
  List<ItemSearchRequest> request,
  @WebParam(name = "OperationRequest",
            targetNamespace =
            "http://webservices.amazon.com/AWSECommerceService/2008-03-03",
            mode = WebParam.Mode.OUT)
  Holder<OperationRequest> operationRequest,
  @WebParam(name = "Items",
            targetNamespace =
            "http://webservices.amazon.com/AWSECommerceService/2008-03-03",
            mode = WebParam.Mode.OUT)
  Holder<List<Items>> items);
```

The last two parameters, named `operationRequest` and `items`, are described as `WebParam.Mode.OUT` to signal that they represent return values from the E-Commerce service to the requester. An `OUT` parameter is returned in a Java `Holder` object. The method `itemSearch` thus reflects the XSD request and response types from the WSDL. The XSD type `ItemSearch`, which is the request wrapper, has eight subelements such as `MarketplaceDomain`, `AWSAccessKeyId`, and `ItemSearchRequest`. Each of these subelements occurs as a parameter in the `itemSearch` method. The XSD type `ItemSearchResponse` has two subelements, named `OperationRequest` and `Items`, which are the last two parameters (the `OUT` parameters) in the `itemSearch` method. The tricky part for the programmer, of course, is that `itemSearch` becomes hard to invoke precisely because of the 10 arguments, especially because the last 2 arguments hold the return values.

Why does the wrapped style result in such a complicated client? Recall that the wrapped style is meant to give a `document`-style service the look and feel of an `rpc`-style service without giving up the advantages of `document` style. The wrapped `document` style requires a wrapper XML `element`, typically with the name of a web service operation such as `ItemSearch`, that has a typed XML `subelement` per operation parameter. In the case of the `itemSearch` operation in the E-Commerce service, there are eight `in` or request parameters and two `out` or response parameters, including the critical `Items` response parameter. The XML subelements that represent the parameters occur in an `xs:sequence`, which means that each parameter is positional. For example, the response parameter `Items` must come last in the list of 10 parameters because the part `@ResponseWrapper` comes after the `@RequestWrapper` and the `Items` parameter comes last in the `@ResponseWrapper`. The upshot is that the wrapped style makes for a very tricky invocation of the `itemSearch` method. The wrapped style, without a workaround, may well produce an irritated programmer. The next section presents a workaround.

An E-Commerce Client in Unwrapped Style

The client built from artifacts in the unwrapped style is simpler than the client built from artifacts in the wrapped style. However, artifacts for the simplified E-Commerce client are generated from the very same WSDL as the artifacts for the more complicated, wrapped-style client. The underlying SOAP messages have the same structure with either the complicated or the simplified client. Yet the clients differ significantly—the simplified client is far easier to code and to understand. Here is the call to `invokeSearch` in the simplified client:

```
ItemSearchResponse response = port.itemSearch(item_search);
```

The `itemSearch` method now takes one argument and returns a value, which is assigned to the object reference `response`. Unlike the complicated client, the simplified client has no `Holder` of a return value, which is an exotic and difficult construct. The invocation of `itemSearch` is familiar, identical in style to method calls in standalone applications. The full simplified client is in Example 2-13.

Example 2-13. An E-Commerce Java client in unwrapped style

```
import awsClient2.AWSECommerceService;
import awsClient2.AWSECommerceServicePortType;
import awsClient2.ItemSearchRequest;
import awsClient2.ItemSearchResponse;
import awsClient2.ItemSearch;
import awsClient2.Items;
import awsClient2.Item;
import java.util.List;

class AmazonClientU { // U is for Unwrapped style
    public static void main(String[ ] args) {
        // Usage
        if (args.length != 1) {
            System.err.println("Usage: java AmazonClientW <access key<");
            return;
        }
        final String access_key = args[0];

        // Create service and get portType reference.
        AWSECommerceService service = new AWSECommerceService();
        AWSECommerceServicePortType port =
          service.getAWSECommerceServicePort();

        // Create request.
        ItemSearchRequest request = new ItemSearchRequest();

        // Add details to request.
        request.setSearchIndex("Books");
        request.setKeywords("quantum gravity");
        ItemSearch item_search= new ItemSearch();
        item_search.setAWSAccessKeyId(access_key);
        item_search.getRequest().add(request);

        // Invoke service operation and get response.
        ItemSearchResponse response = port.itemSearch(item_search);

        List<Items> item_list = response.getItems();
        for (Items next : item_list)
           for (Item item : next.getItem())
              System.out.println(item.getItemAttributes().getTitle());
    }
}
```

Generating the simplified client with the *wsimport* is ironically more complicated than generating the complicated client. Here is the command, on three lines to enhance readability:

```
% wsimport -keep -p awsClient2 \
   http://ecs.amazonaws.com/AWSECommerceService/AWSECommerceService.wsdl \
   -b custom.xml .
```

The -b flag at the end specifies a customized `jaxws:bindings` document, in this case in the file *custom.xml*, that overrides *wsimport* defaults, in this case a `WrapperStyle` setting of `false`. Example 2-14 shows the document with the customized binding information.

Example 2-14. A customized bindings document for wsimport

```
<jaxws:bindings
    wsdlLocation =
      "http://ecs.amazonaws.com/AWSECommerceService/AWSECommerceService.wsdl"
    xmlns:jaxws="http://java.sun.com/xml/ns/jaxws">
  <jaxws:enableWrapperStyle>false</jaxws:enableWrapperStyle>
</jaxws:bindings>
```

The impact of the customized binding document is evident in the generated artifacts. For example, the segment of the `AWSECommerceServicePortType` artifact becomes:

```
@SOAPBinding(parameterStyle = SOAPBinding.ParameterStyle.BARE)
public interface AWSECommerceServicePortType {
  ...
  @WebMethod(operationName = "ItemSearch",
             action = "http://soap.amazon.com")
  @WebResult(name = "ItemSearchResponse",
             targetNamespace =
               "http://webservices.amazon.com/AWSECommerceService/2008-04-07",
             partName = "body")
  public ItemSearchResponse itemSearch(
      @WebParam(name = "ItemSearch",
                targetNamespace =
                  "http://webservices.amazon.com/AWSECommerceService/2008-04-07",
                partName = "body")
      ItemSearch body);
```

The *wsimport*-generated interface `AWSECommerceServicePortType` now has the annotation `@SOAPBinding.ParameterStyle.BARE`, where `BARE` is the alternative to `WRAPPED`. JWS names the attribute `parameterStyle` because the contrast between wrapped and unwrapped in document-style web services comes down to how parameters are represented in the SOAP body. In the unwrapped style, the parameters occur *bare*; that is, as a sequence of unwrapped XML subelements in the SOAP body. In the wrapped style, the parameters occur as *wrapped* XML subelements of an XML element with the name of the service operation; and the wrapper XML element is the only direct subelement of the SOAP body.

What may be surprising is that the structure of the underlying SOAP messages, both the request and the response, remain unchanged. For instance, the request message from the simplified client `AmazonClientU` is identical in structure to the request message from the complicated client `AmazonClientW`. Here is the body of the SOAP envelope from a request that the simplified client generates:

```
<soapenv:Body>
  <ns1:ItemSearch>
    <ns1:AWSAccessKeyId>...</ns1:AWSAccessKeyId>
    <ns1:Request>
```

```
        <ns1:Keywords>quantum gravity</ns1:Keywords>
        <ns1:SearchIndex>Books</ns1:SearchIndex>
      </ns1:Request>
   </ns1:ItemSearch>
   </soapenv:Body>
```

The SOAP body contains a single wrapper element, `ns1:ItemSearch`, with subelements for the access key identifier and the request details. The complicated, wrapped-style client generates requests with the identical structure.

The key difference is in the Java client code, of course. The simplified client, with the unwrapped `parameterStyle`, calls `invokeSearch` with one argument of type `ItemSearch` and expects a single response of type `ItemSearchResponse`. So the `parameterStyle` with a value of `BARE` eliminates the complicated call to `invokeSearch` with 10 arguments, 8 of which are arguments bound to the subelements in the `@RequestWrapper`, and 2 of which are arguments bound to subelements in the `@ResponseWrapper`.

The E-Commerce example shows that the wrapped `document`-style, despite its advantages, can complicate the programming of a client. In that example, the simplified client with the `BARE` or unwrapped `parameterStyle` is a workaround.

The underlying WSDL for the E-Commerce example could be simplified, which would make clients against the service easier to code. Among SOAP-based web services available commercially, it is not unusual to find complicated WSDLs that in turn complicate the clients written against the service.

Tradeoffs Between the RPC and Document Styles

JWS still supports both `rpc` and `document` styles, with `document` as the default; for both styles, only `literal` encoding is supported in compliance with the WS-I Basic Profile. The issue of `rpc` versus `document` often is touted as a *freedom of choice* issue. Nonetheless, it is clear that `document` style, especially in the wrapped flavor, is rapidly gaining mind share. This subsection briefly reviews some tradeoffs between the two styles.

As in any tradeoff scenario, the pros and cons need to be read with a critical eye, especially because a particular point may be cited as both a pro and a con. One familiar complaint against the `rpc` style is that it imposes the request/response pattern on the service. However, this pattern remains the dominant one in real-world, SOAP-based web services, and there are obviously situations (for instance, validating a credit card to be used in a purchase) in which the request/response pattern is needed.

Here are some upsides of the `rpc` style:

- The automatically generated WSDL is relatively short and simple because there is no `types` section.
- Messages in the WSDL carry the names of the underlying web service operations, which are `@WebMethod`s in a Java-based service. The WSDL thus has a *what you see is what you get* style with respect to the service's operations.

- Message throughput may improve because the messages do not carry any type-encoding information.

Here are some downsides to the `rpc` style:

- The WSDL, with its empty `types` section, does not provide an XSD against which the body of a SOAP message can be validated.
- The service cannot use arbitrarily rich data types because there is no XSD to define such types. The service is thus restricted to relatively simple types such as integers, strings, dates, and arrays of such.
- This style, with its obvious link to the request/response pattern, encourages *tight coupling* between the service and the client. For example, the Java client `ch01.ts.TimeClient` *blocks* on the call:

      port.getTimeAsString()

  until either the service responds or an exception is thrown. This same point is sometimes made by noting that the `rpc` style has an inherently *synchronous* as opposed to *asynchronous* invocation idiom. The next section offers a workaround, which shows how JWS supports nonblocking clients under the request/response pattern.
- Java services written in this style may not be consumable in other frameworks, thus undermining interoperability. Further, long-term support within the web services community and from the WS-I group is doubtful.

Here are some upsides of the `document` style:

- The body of a SOAP message can be validated against the XSD in the `types` section of the WSDL.
- A service in this style can use arbitrarily rich data types, as the XML Schema language supports not only simple types such as integers, strings, and dates, but also arbitrarily rich complex types.
- There is great flexibility in how the body of a SOAP message is structured so long as the structure is clearly defined in an XSD.
- The *wrapped* convention provides a way to enjoy a key upside of the `rpc` style—naming the body of a SOAP message after the corresponding service operation—without enduring the downsides of the `rpc` style.

Here are some downsides of the `document` style:

- In the unwrapped variant, the SOAP message does not carry the name of the service operation, which can complicate the dispatching of messages to the appropriate program code.
- The wrapped variant adds a level of complexity, in particular at the API level. Writing a client against a wrapped-`document` service can be challenging, as the `AmazonClientW` example shows.

- The wrapped variant does not support overloaded service operations because the XML wrapper `element` in the body of a SOAP message must have the name of the service operation. In effect, then, there can be only one operation for a given `element` name.

An Asynchronous E-Commerce Client

As noted earlier, the original `TimeClient` blocks on calls against the `TimeServer` service operations. For example, the call:

```
port.getTimeAsString()
```

blocks until either a response from the web service occurs or an exception is thrown. The call to `getTimeAsString` is, therefore, known as a *blocking* or *synchronous* call. JWS also supports *nonblocking* or *asynchronous* clients against web services.

Example 2-15 shows a client that makes an asynchronous call against the E-Commerce service.

Example 2-15. An asynchronous client against the E-Commerce service

```
import javax.xml.ws.AsyncHandler;
import javax.xml.ws.Response;

import awsClient3.AWSECommerceService;
import awsClient3.AWSECommerceServicePortType;
import awsClient3.ItemSearchRequest;
import awsClient3.ItemSearchResponse;
import awsClient3.ItemSearch;
import awsClient3.Items;
import awsClient3.Item;

import java.util.List;
import java.util.concurrent.ExecutionException;

class AmazonAsyncClient {
    public static void main(String[ ] args) {
        // Usage
        if (args.length != 1) {
            System.err.println("Usage: java AmazonAsyncClient <access key>");
            return;
        }
        final String access_key = args[0];

        // Create service and get portType reference.
        AWSECommerceService service = new AWSECommerceService();
        AWSECommerceServicePortType port = service.getAWSECommerceServicePort();

        // Create request.
        ItemSearchRequest request = new ItemSearchRequest();

        // Add details to request.
        request.setSearchIndex("Books");
```

```
        request.setKeywords("quantum gravity");
        ItemSearch item_search= new ItemSearch();
        item_search.setAWSAccessKeyId(access_key);
        item_search.getRequest().add(request);

        port.itemSearchAsync(item_search, new MyHandler());

        // In this case, just sleep to give the search process time.
        // In a production application, other useful tasks could be
        // performed and the application could run indefinitely.
        try {
            Thread.sleep(400);
        }
        catch(InterruptedException e) { System.err.println(e); }
    }

    // The handler class implements handleResponse, which executes
    // if and when there's a response.
    static class MyHandler implements AsyncHandler<ItemSearchResponse> {
        public void handleResponse(Response<ItemSearchResponse> future) {
            try {
                ItemSearchResponse response = future.get();
                List<Items> item_list = response.getItems();
                for (Items next : item_list)
                    for (Item item : next.getItem())
                        System.out.println(item.getItemAttributes().getTitle());
            }
            catch(InterruptedException e) { System.err.println(e); }
            catch(ExecutionException e) { System.err.println(e); }
        }
    }
}
```

The nonblocking E-Commerce client uses artifacts generated by *wsimport*, again with a customized bindings file. Here is the file:

```
<jaxws:bindings
    wsdlLocation=
      "http://ecs.amazonaws.com/AWSECommerceService/AWSECommerceService.wsdl"
    xmlns:jaxws="http://java.sun.com/xml/ns/jaxws">
  <jaxws:enableWrapperStyle>false</jaxws:enableWrapperStyle>
  <jaxws:enableAsyncMapping>true</jaxws:enableAsyncMapping>
</jaxws:bindings>
```

with the enableAsyncMapping attribute set to true.

The nonblocking call can follow different styles. In the style shown here, the call to itemSearchAsync takes two arguments: the first is an ItemSearchRequest, and the second is a class that implements the AsyncHandler interface, which declares a single method named handleResponse. The call is:

```
port.itemSearchAsync(item_search, new MyHandler());
```

If and when an `ItemSearchResponse` comes from the E-Commerce service, the method `handleResponse` in the `MyHandler` class executes as a separate thread and prints out the books' titles.

There is a version of `itemSearchAsync` that takes one argument, an `ItemSearchRequest`. In this version the call also returns if and when the E-Commerce service sends a response, which then can be processed as in the other E-Commerce clients. In this style, the application might start a separate thread to execute this code segment:

```
Response<ItemSearchResponse> res = port.itemSearchAsync(item_search);
try {
    ItemSearchResponse response = res.get();
    List<Items> item_list = response.getItems();
    for (Items next : item_list)
        for (Item item : next.getItem())
            System.out.println(item.getItemAttributes().getTitle());
}
catch(InterruptedException e) { System.err.println(e); }
catch(ExecutionException e) { System.err.println(e); }
```

JWS is flexible in supporting nonblocking as well as the default blocking clients. In the end, it is application logic that determines which type of client is suitable.

The wsgen Utility and JAX-B Artifacts

Any document-style service, wrapped or unwrapped, requires the kind of artifacts that the *wsgen* utility produces. It is time to look again at this utility. A simple experiment underscores the role that the utility plays. To begin, the line:

```
@SOAPBinding(style = Style.RPC)
```

should be commented out or deleted from the SEI `ch01.ts.TimeServer`. With this annotation gone, the web service becomes document style rather than rpc style. After recompiling the altered SEI, try to publish the service with the command:

```
% java ch01.ts.TimeServerPublisher
```

The resulting error message is:

```
Exception in thread "main" com.sun.xml.internal.ws.model.RuntimeModelerException:
runtime modeler error: Wrapper class ch01.ts.jaxws.GetTimeAsString is not found.
```

The message is obscure in citing the immediate problem rather than the underlying cause. The immediate problem is that the publisher cannot find the class `ch01.ts.jaxws.GetTimeAsString`. Indeed, at this point the package `ch01.ts.jaxws` that houses the class does not even exist. The publisher cannot generate the WSDL because the publisher needs Java classes such as `GetTimeAsString` to do so. The *wsgen* utility produces the classes required to build the WSDL, classes known as *wsgen* artifacts. The command:

```
% wsgen -keep -cp . ch01.ts.TimeServerImpl
```

produces the artifacts and, if necessary, the package `ch01.ts.jaxws` that houses these artifacts. In the `TimeServer` example, there are four messages altogether: the request and response messages for the `getTimeAsString` operation, and the request and response messages for the `getTimeAsElapsed` operation. The *wsgen* utility generates a Java class—hence, a Java data type—for each of these messages. It is these Java types that the publisher uses to generate the WSDL for a `document`-style service. So each of the Java data types is bound to an XML Schema type, which serves as a type for one of the four messages involved in the service. Considered from the other direction, a `document`-style web service has *typed* messages. The *wsgen* artifacts are the Java types from which the XML Schema types for the messages are derived.

A SOAP-based web service, in either `document` or `rpc` style, should be interoperable: a client application written in one language should be able to interact seamlessly with a service written in another despite any differences in data types between the two languages. A shared type system such as XML Schema is, therefore, the key to interoperability. The `document` style extends typing to the service messages; and typed messages requires the explicit binding of Java and XML Schema types.

Any `document`-style service, wrapped or unwrapped, has a WSDL with an XSD in its `types` section. (The term *in* is being used loosely here. The `types` section could import the XSD or link to the XSD.) The types in the WSDL's associated XSD bind to types in a service-implementation or client language such as Java. The *wsgen* utility, introduced earlier to change the `TimeServer` service from `rpc` to `document` style, generates Java types that bind to XML Schema types. Under the hood, this utility uses the packages associated with *JAX-B* (Java API for **X**ML-Binding). In a nutshell, JAX-B supports conversions between Java and XML types.

A JAX-B Example

Before looking again at the artifacts that *wsgen* generates, let's consider a JAX-B example to get a better sense of what is going on in *wsgen*. Example 2-16 is the code for a programmer-defined Java type, the `Person` class, and Example 2-17 is another programmer-defined Java type, the `Skier` class. Each class declaration begins with a single annotation for Java-to-XML binding. The `Person` class is annotated as an `@XmlType`, whereas the `Skier` class is annotated as an `@XmlRootElement`. At the design level, a `Skier` is also a `Person`; hence, the `Skier` class has a `Person` field.

Example 2-16. The Person class for a JAX-B example

```
import javax.xml.bind.annotation.XmlType;

@XmlType
public class Person  {
    // fields
    private String name;
    private int    age;
    private String gender;
```

```
    // constructors
    public Person() { }

    public Person(String name, int age, String gender){
        setName(name);
        setAge(age);
        setGender(gender);
    }

    // properties: name, age, gender
    public String getName() { return name; }
    public void setName(String name) { this.name = name; }

    public int getAge() { return age;  }
    public void setAge(int age) { this.age = age; }

    public String getGender() { return gender; }
    public void setGender(String gender) { this.gender = gender; }
}
```

Example 2-17. The Skier class for a JAX-B example

```
import javax.xml.bind.annotation.XmlRootElement;
import java.util.Collection;

@XmlRootElement
public class Skier  {
    // fields
    private Person person;
    private String national_team;
    private Collection major_achievements;
    // constructors
    public Skier() { }
    public Skier (Person person,
                  String national_team,
                  Collection<String> major_achievements) {
        setPerson(person);
        setNationalTeam(national_team);
        setMajorAchievements(major_achievements);
    }
    // properties
    public Person getPerson() { return person; }
    public void setPerson (Person person) { this.person = person; }

    public String getNationalTeam() { return national_team; }
    public void setNationalTeam(String national_team) {
        this.national_team = national_team;
    }

    public Collection getMajorAchievements() { return major_achievements; }
    public void setMajorAchievements(Collection major_achievements) {
        this.major_achievements = major_achievements;
    }
}
```

The @XmlType and @XmlRootElement annotations direct the marshaling of Skier objects, where *marshaling* is the process of encoding an in-memory object (for example, a Skier) as an XML document so that, for instance, the encoding can be sent across a network to be *unmarshaled* or decoded at the other end. In common usage, the distinction between *marshal* and *unmarshal* is very close and perhaps identical to the *serialize/deserialize* distinction. I use the distinctions interchangeably. JAX-B supports the marshaling of in-memory Java objects to XML documents and the unmarshaling of XML documents to in-memory Java objects.

The annotation @XmlType, applied in this example to the Person class, indicates that JAX-B should generate an XML Schema type from the Java type. The annotation @XmlRootElement, applied in this example to the Skier class, indicates that JAX-B should generate an XML *document* (outermost or root) element from the Java class. Accordingly, the resulting XML in this example is a document whose outermost element represents a skier; and the document has a nested element that represents a person.

The Marshal application in Example 2-18 illustrates marshaling and unmarshaling.

Example 2-18. The Marshal application for a JAX-B example

```java
import java.io.File;
import java.io.OutputStream;
import java.io.FileOutputStream;
import java.io.InputStream;
import java.io.FileInputStream;
import java.io.IOException;
import javax.xml.bind.JAXBContext;
import javax.xml.bind.Marshaller;
import javax.xml.bind.Unmarshaller;
import javax.xml.bind.JAXBException;
import java.util.List;
import java.util.ArrayList;

class Marshal {
    private static final String file_name = "bd.mar";
    public static void main(String[ ] args) {
        new Marshal().run_example();
    }

    private void run_example() {
        try {
            JAXBContext ctx = JAXBContext.newInstance(Skier.class);
            Marshaller m = ctx.createMarshaller();
            m.setProperty(Marshaller.JAXB_FORMATTED_OUTPUT, true);

            // Marshal a Skier object: 1st to stdout, 2nd to file
            Skier skier = createSkier();
            m.marshal(skier, System.out);

            FileOutputStream out = new FileOutputStream(file_name);
            m.marshal(skier, out);
            out.close();
```

```
        // Unmarshal as proof of concept
        Unmarshaller u = ctx.createUnmarshaller();
        Skier bd_clone = (Skier) u.unmarshal(new File(file_name));
        System.out.println();
        m.marshal(bd_clone, System.out);
    }
    catch(JAXBException e) { System.err.println(e); }
    catch(IOException e) { System.err.println(e); }
}

private Skier createSkier() {
    Person bd = new Person("Bjoern Daehlie", 41, "Male");
    List<String> list = new ArrayList<String>();
    list.add("12 Olympic Medals");
    list.add("9 World Championships");
    list.add("Winningest Winter Olympian");
    list.add("Greatest Nordic Skier");
    return new Skier(bd, "Norway", list);
}
}
```

The application constructs a `Skier` object, including a `Person` object encapsulated as a `Skier` field, and sets the appropriate `Skier` and `Person` properties. A critical feature of marshaling is that the process preserves an object's state in the encoding, where an object's *state* comprises the values of its instance (that is, non-`static`) fields. In the case of a `Skier` object, the marshaling must preserve state information such as the skier's major accomplishments together with `Person`-specific state information such as the skier's name and age. The `@XmlRootElement` annotation in the declaration of the `Skier` class directs the marshaling as follows: the `Skier` object is encoded as an XML document whose outermost (that is, *document*) element is named `skier`. The `@XmlType` annotation in the `Person` class directs the marshaling as follows: the `skier` XML document has a `person` subelement, which in turn has subelements for the `name`, `age`, and `gender` properties of the marshaled skier.

By default, JAX-B marshaling follows standard Java and JavaBean naming conventions. For example, the Java class `Skier` becomes the XML tag named `skier` and the Java class `Person` becomes the XML tag named `person`. For both the `Skier` and the encapsulated `Person` objects, the JavaBean-style `get` methods are invoked (e.g., `getNationalTeam` in `Skier` and `getAge` in `Person`) to populate the XML document with state information about the skier.

It is possible to override, with annotations, JavaBean naming conventions in marshaling and unmarshaling classes. Following, for example, is a *wsgen*-generated artifact with the `get` and `set` methods that do not follow JavaBean naming conventions. The annotation of interest is in bold:

```
...
@XmlRootElement(name = "getTimeAsElapsedResponse", namespace = "http://ts.ch02/")
@XmlAccessorType(XmlAccessType.FIELD)
@XmlType(name = "getTimeAsElapsedResponse", namespace = "http://ts.ch02/")
```

```
public class GetTimeAsElapsedResponse {
    @XmlElement(name = "return", namespace = "")
    private long _return;
    public long get_return() { return this._return; }

    public void set_return(long _return) { this._return = _return; }
}
```

The annotation in bold indicates that the field named **_return** will be marshaled and unmarshaled rather than a property defined with a **get/set** method pair that follows the usual JavaBean naming conventions.

Other annotation attributes can be set to override the default naming conventions. For example, if the annotation in the **Skier** class declaration is changed to:

```
@XmlRootElement(name = "NordicSkier")
```

then the resulting XML document begins:

```
<?xml version="1.0" encoding="UTF-8" standalone="yes"?>
<NordicSkier>
```

Once the **Skier** object has been constructed, the **Marshal** application marshals the object to the standard output and, to set up unmarshaling, to a local file. Here is the XML document that results from marshaling the legendary Nordic skier Bjoern Daehlie:

```
<?xml version="1.0" encoding="UTF-8" standalone="yes"?>
<skier>
    <majorAchievements
        xmlns:xsi="http://www.w3.org/2001/XMLSchema-instance"
        xmlns:xs="http://www.w3.org/2001/XMLSchema"
        xsi:type="xs:string">
    12 Olympic Medals
    </majorAchievements>
    <majorAchievements
        xmlns:xsi="http://www.w3.org/2001/XMLSchema-instance"
        xmlns:xs="http://www.w3.org/2001/XMLSchema"
        xsi:type="xs:string">
    9 World Championships
    </majorAchievements>
    <majorAchievements
        xmlns:xsi="http://www.w3.org/2001/XMLSchema-instance"
        xmlns:xs="http://www.w3.org/2001/XMLSchema"
        xsi:type="xs:string">
    Winningest Winter Olympian
    </majorAchievements>
    <majorAchievements
        xmlns:xsi="http://www.w3.org/2001/XMLSchema-instance"
        xmlns:xs="http://www.w3.org/2001/XMLSchema"
        xsi:type="xs:string">
    Greatest Nordic Skier
    </majorAchievements>
    <nationalTeam>Norway</nationalTeam>
    <person>
        <age>41</age>
        <gender>Male</gender>
```

```
        <name>Bjoern Daehlie</name>
      </person>
  </skier>
```

The unmarshaling process constructs a `Skier`, with its encapsulated `Person` field, from the XML document. JAX-B unmarshaling requires that each class have a `public` no-argument constructor, which is used in the construction. After the object is constructed, the appropriate `set` methods are invoked (for instance, `setNationalTeam` in `Skier` and `setAge` in `Person`) to restore the marshaled skier's state. The marshaling and unmarshaling processes are remarkably clear and straightforward at the code level.

Marshaling and wsgen Artifacts

Now we can tie the *wsgen* utility and marshaling together. Recall that there were two steps to changing the `TimeServer` application from `rpc` to `document` style. First, the annotation:

```
@SOAPBinding(style = Style.RPC)
```

is commented out in the SEI, `ch01.ts.TimeServer`. Second, the *wsgen* utility is executed in the working directory against the `ch01.ts.TimeServerImpl` class:

```
% wsgen -keep -cp . ch01.ts.TimeServerImpl
```

The *wsgen* invocation generates two source and two compiled files in the automatically created subdirectory *ch01/ts/jaxws*. Example 2-19 shows the *wsgen* artifact, the class named `GetTimeAsElapsedResponse`, with the comments removed.

Example 2-19. A Java class generated with wsgen

```java
package ch01.ts.jaxws;

import javax.xml.bind.annotation.XmlAccessorType;
import javax.xml.bind.annotation.XmlElement;
import javax.xml.bind.annotation.XmlRootElement;
import javax.xml.bind.annotation.XmlType;

@XmlRootElement(name = "getTimeAsElapsedResponse", namespace = "http://ts.ch01/")
@XmlAccessorType(XmlAccessType.FIELD)
@XmlType(name = "getTimeAsElapsedResponse", namespace = "http://ts.ch01/")
public class GetTimeAsElapsedResponse {
    @XmlElement(name = "return", namespace = "")
    private long _return;
    public long get_return() { return this._return; }
    public void set_return(long _return) { this._return = _return; }
}
```

Of particular interest is the `@XmlType` annotation applied to the class. The annotation sets the `name` attribute to `getTimeAsElapsedResponse`, the name of a SOAP message returned from the web service to the client on a call to the `getTimeAsElapsed` operation. The `@XmlType` annotation means that such SOAP response messages are *typed*; that is, that they satisfy an XML Schema. In a document-style service, a SOAP message is typed.

Example 2-20 shows a modified version of the Marshal application, renamed MarshalGTER to signal that the revised application is to be run against the *wsgen*-generated class named GetTimeAsElapsedResponse.

Example 2-20. Code to illustrate the link between wsgen and JAX-B

```
import javax.xml.bind.JAXBContext;
import javax.xml.bind.Marshaller;
import javax.xml.bind.JAXBException;

import ch01.ts.jaxws.GetTimeAsElapsedResponse;

class MarshalGTER {
    private static final String file_name = "gter.mar";

    public static void main(String[ ] args) {
        new MarshalGTER().run_example();
    }

    private void run_example() {
        try {
            JAXBContext ctx =
                JAXBContext.newInstance(GetTimeAsElapsedResponse.class);
            Marshaller m = ctx.createMarshaller();
            m.setProperty(Marshaller.JAXB_FORMATTED_OUTPUT, true);

            GetTimeAsElapsedResponse tr = new GetTimeAsElapsedResponse();
            tr.set_return(new java.util.Date().getTime());

            m.marshal(tr, System.out);
        }
        catch(JAXBException e) { System.err.println(e); }
    }
}
```

The marshaled XML document on a sample run was:

```
<?xml version="1.0" encoding="UTF-8" standalone="yes"?>
<ns2:getTimeAsElapsedResponse xmlns:ns2="http://ts.ch01/">
    <return>1209174518855</return>
</ns2:getTimeAsElapsedResponse>
```

For reference, here is the response message from the document-style version of the TimeServer service:

```
<?xml version="1.0" ?>
<soapenv:Envelope
    xmlns:soapenv="http://schemas.xmlsoap.org/soap/envelope/"
    xmlns:xsd="http://www.w3.org/2001/XMLSchema"
    xmlns:ns1="http://ts.ch01/">
  <soapenv:Body>
    <ns1:getTimeAsElapsedResponse>
```

```
            <return>1209181713849</return>
          </ns1:getTimeAsElapsedResponse>
        </soapenv:Body>
      </soapenv:Envelope>
```

There are some incidental differences between the body of the SOAP message and the output of the `MarshalGTER` application. For instance, the namespace prefix is `ns1` in the SOAP message but `ns2` in the marshaled XML document. Note, however, that the all-important namespace URI is the same in both: `http://ts.ch01`. In the SOAP message, the namespace prefix is defined in the `Envelope` element rather than in the element `getTimeAsElapsedResponse`. Of interest here is that the *wsgen*-generated artifacts such as the `GetTimeAsElapsedResponse` class provide the annotated types that the underlying SOAP libraries can use to marshal Java objects of some type into XML documents and to unmarshal such documents into Java objects of the appropriate type.

An Overview of Java Types and XML Schema Types

Java's primitive types such as `int` and `byte` bind to similarly named XML Schema types, in this case `xsd:int` and `xsd:byte`, respectively. The `java.lang.String` type binds to `xsd:string`, and the `java.util.Calendar` typebinds to each of `xsd:date`, `xsd:time`, and `xsd:dateTime`. The XML Schema type `xsd:decimal` binds to the Java type `BigDecimal`. Not all bindings are obvious and the same Java type—for example, `int` may match up with several XSD types, for instance, `xsd:int` and `xsd:unsignedShort`. Here is the reason for the apparent mismatch between a Java `int` and an XSD `xsd:unsignedShort`. Java technically does not have unsigned integer types. (You could argue, of course, that a Java `char` is really an unsigned 16-bit integer.) The maximam value of the Java 16-bit short integer is 32,767, but the maximum value of the `xsd:unsignedShort` integer is 65,535, a value within the range of a 32-bit Java `int`.

What JAX-B brings to the table is a framework for binding arbitrarily rich Java types, with the `Skier` class as but a hint of the possibilities, to XML types. Instances of these Java types can be marshaled to XML document instances, which in turn can be un-marshaled to instances of either Java types or types in some other language.

How the *wsgen* utility and the JAX-B packages interact in JWS now can be summarized. A Java web service in `document` rather than `rpc` style has a nonempty `types` section in its WSDL. This section defines, in the XML Schema language, the types required for the web service. The *wsgen* utility generates, from the SIB, Java classes that are coun-terparts of XSD types. These *wsgen* artifacts are available for the underlying JWS li-braries, in particular for the JAX-B family of packages, to convert (marshal) instances of Java types (that is, Java in-memory objects) into XML instances of XML types (that is, into XML document instances that satisfy an XML Schema document). The inverse operation is used to convert (unmarshal) an XML document instance to an in-memory object, an object of a Java type or a comparable type in some other language. The *wsgen* utility thus produces the artifacts that support interoperability for a Java-based web service. The JAX-B libraries provide the under-the-hood support to convert between Java and XSD types. For the most part, the *wsgen* utility can be used without our bothering to inspect the artifacts that it produces. For the most part, JAX-B remains unseen infrastructure.

Beyond the Client-Side wsgen

A web service's SEI contains all of the information required to generate the *wsgen* artifacts. After all, the SEI declares the service operations as `@WebMethod`s and these declarations specify argument and return types.

In the current Metro release, the `Endpoint` publisher automatically generates the *wsgen* artifacts if the programmer does not. For example, once the sample service has been compiled, the command:

```
% java -cp .:$METRO_HOME/lib/jaxws-rt.jar test.WS1Pub
```

publishes the web service even though no *wsgen* artifacts were generated beforehand. (Under Windows, `$METRO_HOME` becomes `%METRO_HOME%`.) Here is the output:

```
com.sun.xml.ws.model.RuntimeModeler getRequestWrapperClass
INFO: Dynamically creating request wrapper Class test.jaxws.Op
com.sun.xml.ws.model.WrapperBeanGenerator createBeanImage
INFO:
@XmlRootElement(name=op, namespace=http://test/)
@XmlType(name=op, namespace=http://test/)
public class test.jaxws.Op {
    @XmlRootElement(name=arg0, namespace=)
    public I arg0
}
com.sun.xml.ws.model.RuntimeModeler getResponseWrapperClass
INFO: Dynamically creating response wrapper bean Class
      test.jaxws.OpResponse
com.sun.xml.ws.model.WrapperBeanGenerator createBeanImage
INFO:
@XmlRootElement(name=opResponse, namespace=http://test/)
@XmlType(name=opResponse, namespace=http://test/)
public class test.jaxws.OpResponse {
    @XmlRootElement(name=return, namespace=)
    public I _return
}
```

In time this convenient feature of the Metro release will make its way into core Java so that the *wsgen* step in document-style services can be avoided. It is still helpful to understand exactly how the JAX-B artifacts from the *wsgen* utility figure in Java-based web services.

Generating a WSDL with the wsgen Utility

The *wsgen* utility also can be used to generate a WSDL document for a web service. For example, the command:

```
% wsgen -cp "." -wsdl ch01.ts.TimeServerImpl
```

generates a WSDL for the original `TimeServer` service. The `TimeServer` service is `rpc` rather than `document` style, which is reflected in the WSDL. There is one critical difference between the WSDL generated with the *wsgen* utility and the one retrieved at runtime after the web service has been published: the *wsgen*-generated WSDL does not include the service endpoint, as this URL depends on the actual publication of the service. Here is the relevant segment from the *wsgen*-generated WSDL:

```
<service name="TimeServerImplService">
  <port name="TimeServerImplPort" binding="tns:TimeServerImplPortBinding">
    <soap:address location="REPLACE_WITH_ACTUAL_URL"/>
  </port>
</service>
```

Except for this difference, the two WSDLs are the same service contract.

Generating a WSDL directly from *wsgen* will be useful in later security examples. If a web service is secured, then so is its WSDL; hence, the first step in writing a *wsimport*-supported client to test the service is to access the WSDL, and *wsgen* is an easy way to take this step.

WSDL Wrap-Up

This section completes the chapter with a look at two issues. The first issue is whether the web service code—in particular, its SEI and SIB—should come before or after the WSDL. In other words, should the WSDL be generated automatically from the service implementation or should the WSDL be designed and written *before* the web service code is written? This issue has been popularized as the *code first* versus the *contract first* controversy. The second issue is about the limited information that a WSDL provides to potential consumers of the corresponding web service.

Code First Versus Contract First

The issue can be expressed as a question: should the web service code be used to generate the WSDL automatically or should the WSDL, independently designed and created, be used to guide web service coding? All of the examples so far take the *code-first* approach because of its most obvious advantage: it is easy. However, there are some obvious shortcomings and even dangers in this approach, including:

- If the service changes, the WSDL automatically changes. The WSDL thereby loses some of its appeal for creating client artifacts, as this process may have to be repeated over and over. One of the first principles of software development is that an interface, once published, should be regarded as immutable so that code once written against a published interface never has to be rewritten. The *code-first* approach may compromise this principle.

- The *code-first* approach usually results in a service contract that provides few, if any, provisions for handling tricky but common problems in distributed systems such as partial failure of the service. The reason is that the service may be programmed as if it were a standalone application rather than part of a distributed application with consumers on other hosts.

- If the service implementation is complicated or even messy, these features carry over into a WSDL that may be difficult to understand and to use in the generation of client artifacts. In short, the *code-first* approach is clearly not consumer-oriented.

- The *code-first* approach seems to go against the *language-neutral* theme of SOAP-based web services. If the contract is done first, then the implementation language remains open.

The list could go on and on. My aim, however, is not to jump into the fray but only to caution that the issue has merit. The core difficulty for the *contract-first* advocates is real-world programming practice. A programmer under the pressure of deadlines and application expectations is unlikely to have the time, let alone the inclination, to master the subtleties of WSDLs and then to produce a service contract that suitably guides the programming of the service.

Even the WSDL generated for the simple `TimeServer` web service could use improvement. For example, here is a segment from the XML Schema for the `document`-style `TimeServer` web service:

```
<xs:complexType name="getTimeAsStringResponse">
  <xs:sequence>
    <xs:element name="return" type="xs:string" minOccurs="0"/>
  </xs:sequence>
</xs:complexType>
```

Note that the element for the returned value sets the minimum occurrences to 0 instead of 1. The reason is that the Java-based web service might return `null`, a valid value for a `String` type. To tighten up the service contract by requiring that a non-null string be returned, the 0 can be replaced with 1. (The `minOccurs="0"` also could be deleted because the default value for `minOccurs` and `maxOccurs` is 1.)

A Contract-First Example with wsimport

The *wsimport* utility can be used in support of a contract-first approach. Example 2-21 shows a web service for temperature conversion written in C#.

Example 2-21. A SOAP-based service in C#

```
using System;
using System.Linq;
using System.Web;
using System.Web.Services;
using System.Web.Services.Protocols;
using System.Xml.Linq;

[WebService(Namespace = "http://tempConvertURI.org/")]
[WebServiceBinding(ConformsTo = WsiProfiles.BasicProfile1_1)]
```

```
public class Service : System.Web.Services.WebService {
    public Service () { }

    [WebMethod]
    public double c2f(double t) { return 32.0 + (t * 9.0 / 5.0); }
    [WebMethod]
    public double f2c(double t) { return (5.0 / 9.0) * (t - 32.0); }
}
```

Example 2-22 shows the WSDL for this C# web service.

Example 2-22. The WSDL for the C# service

```
<?xml version="1.0" encoding="utf-8"?>
<wsdl:definitions xmlns:soap="http://schemas.xmlsoap.org/wsdl/soap/"
                  xmlns:tm="http://microsoft.com/wsdl/mime/textMatching/"
                  xmlns:soapenc="http://schemas.xmlsoap.org/soap/encoding/"
                  xmlns:mime="http://schemas.xmlsoap.org/wsdl/mime/"
                  xmlns:tns="http://tempConvertURI.org/"
                  xmlns:s="http://www.w3.org/2001/XMLSchema"
                  xmlns:soap12="http://schemas.xmlsoap.org/wsdl/soap12/"
                  xmlns:http="http://schemas.xmlsoap.org/wsdl/http/"
                  targetNamespace="http://tempConvertURI.org/"
                  xmlns:wsdl="http://schemas.xmlsoap.org/wsdl/">
  <wsdl:types>
    <s:schema elementFormDefault="qualified"
              targetNamespace="http://tempConvertURI.org/">
      <s:element name="c2f">
        <s:complexType>
          <s:sequence>
            <s:element minOccurs="1" maxOccurs="1" name="t" type="s:double"/>
          </s:sequence>
        </s:complexType>
      </s:element>
      <s:element name="c2fResponse">
        <s:complexType>
          <s:sequence>
            <s:element minOccurs="1" maxOccurs="1" name="c2fResult" type="s:double"/>
          </s:sequence>
        </s:complexType>
      </s:element>
      <s:element name="f2c">
        <s:complexType>
          <s:sequence>
            <s:element minOccurs="1" maxOccurs="1" name="t" type="s:double"/>
          </s:sequence>
        </s:complexType>
      </s:element>
      <s:element name="f2cResponse">
        <s:complexType>
          <s:sequence>
            <s:element minOccurs="1" maxOccurs="1" name="f2cResult" type="s:double"/>
          </s:sequence>
        </s:complexType>
      </s:element>
```

```
      </s:schema>
    </wsdl:types>
    <wsdl:message name="c2fSoapIn">
      <wsdl:part name="parameters" element="tns:c2f"/>
    </wsdl:message>
    <wsdl:message name="c2fSoapOut">
      <wsdl:part name="parameters" element="tns:c2fResponse"/>
    </wsdl:message>
    <wsdl:message name="f2cSoapIn">
      <wsdl:part name="parameters" element="tns:f2c"/>
    </wsdl:message>
    <wsdl:message name="f2cSoapOut">
      <wsdl:part name="parameters" element="tns:f2cResponse"/>
    </wsdl:message>
    <wsdl:portType name="ServiceSoap">
      <wsdl:operation name="c2f">
        <wsdl:input message="tns:c2fSoapIn"/>
        <wsdl:output message="tns:c2fSoapOut"/>
      </wsdl:operation>
      <wsdl:operation name="f2c">
        <wsdl:input message="tns:f2cSoapIn"/>
        <wsdl:output message="tns:f2cSoapOut"/>
      </wsdl:operation>
    </wsdl:portType>
    <wsdl:binding name="ServiceSoap" type="tns:ServiceSoap">
      <soap:binding transport="http://schemas.xmlsoap.org/soap/http"/>
      <wsdl:operation name="c2f">
        <soap:operation soapAction="http://tempConvertURI.org/c2f" style="document"/>
        <wsdl:input>
          <soap:body use="literal"/>
        </wsdl:input>
        <wsdl:output>
          <soap:body use="literal"/>
        </wsdl:output>
      </wsdl:operation>
      <wsdl:operation name="f2c">
        <soap:operation soapAction="http://tempConvertURI.org/f2c" style="document"/>
        <wsdl:input>
          <soap:body use="literal"/>
        </wsdl:input>
        <wsdl:output>
          <soap:body use="literal"/>
        </wsdl:output>
      </wsdl:operation>
    </wsdl:binding>
    <wsdl:binding name="ServiceSoap12" type="tns:ServiceSoap">
      <soap12:binding transport="http://schemas.xmlsoap.org/soap/http"/>
      <wsdl:operation name="c2f">
        <soap12:operation soapAction="http://tempConvertURI.org/c2f" style="document"/>
        <wsdl:input>
          <soap12:body use="literal"/>
        </wsdl:input>
        <wsdl:output>
          <soap12:body use="literal"/>
        </wsdl:output>
```

```
      </wsdl:operation>
      <wsdl:operation name="f2c">
        <soap12:operation soapAction="http://tempConvertURI.org/f2c" style="document"/>
        <wsdl:input>
          <soap12:body use="literal"/>
        </wsdl:input>
        <wsdl:output>
          <soap12:body use="literal"/>
        </wsdl:output>
      </wsdl:operation>
    </wsdl:binding>
    <wsdl:service name="Service">
      <wsdl:port name="ServiceSoap" binding="tns:ServiceSoap">
        <soap:address location="http://localhost:1443/TempConvert/Service.asmx"/>
      </wsdl:port>
      <wsdl:port name="ServiceSoap12" binding="tns:ServiceSoap12">
        <soap12:address location="http://localhost:1443/TempConvert/Service.asmx"/>
      </wsdl:port>
    </wsdl:service>
</wsdl:definitions>
```

This WSDL has some features worth noting. The service section (shown in bold) has two ports, one for a SOAP 1.1 version of the service and another for the SOAP 1.2 version. However, the location attributes for the two ports have the same value, which means that the same implementation is available at the two ports. The reason is found in the binding (that is, implementation) section. There are several ways in which two port elements, each a combination of a portType and a binding, could differ. The bindings could differ on the transport attribute with, for example, one binding delivering SOAP over HTTP and the other delivering SOAP over SMTP. The bindings also could differ, as in this example, on the SOAP version: the first is the SOAP 1.1 binding and the second is the SOAP 1.2 binding. However, the service is simple enough that no differences arise because of the SOAP 1.1 versus SOAP 1.2 binding. The two bindings and, therefore, the two ports are identical in this case.

The *wsimport* utility can be used with the WSDL for the C# service to generate the usual artifacts that aid in writing a client. Here is the command:

```
% wsimport -p tcClient -extension  http://localhost:1443/TempConvert/Service.asmx?wsdl
```

The -extension flag is used because the WSDL includes a SOAP 1.2 binding. With these artifacts, the sample client code is straightforward:

```
import tcClient.Service;
import tcClient.ServiceSoap; // port

// A sample Java client against a C# web service
class ClientDotNet {
    public static void main(String[ ] args) {
        Service service = new Service();
        // There's also a getServiceSoap12 for the SOAP 1.2 binding
        ServiceSoap port = service.getServiceSoap();
```

```
        double temp = 98.7;
        System.out.println(port.c2F(temp)); // 209.65999450683594
        System.out.println(port.f2C(temp)); //  37.05555386013455
    }
}
```

However, the main point of this section is to show how the same WSDL also can be used to generate a Java version of the service itself.

Although the WSDL for the C# web service is generated automatically, the origin of the WSDL is unimportant in this example. Further, the WSDL could be edited to make whatever changes seem appropriate and perhaps even rewritten from scratch. The point of interest is that the WSDL is language-neutral; hence, the service that the WSDL describes can be implemented in Java rather than C#. For the sake of the example, the critical point is that the WSDL does not come from Java.

Several details in this WSDL do not make sense for a Java implementation of the service described therein. For one, the service location ends with *Service.asmx*, the file name of the C# implementation. This detail might be changed in the WSDL. Other changes might be made as well; for instance, the service's **port** is currently **ServiceSoap** instead of something more suggestive such as **TempConvert**. In this example, however, the WSDL will be left as is and changes will be made instead to the *wsimport*-generated Java code.

Assuming this time that the WDSL is in the local file named *tempc.wsdl*, the command:

```
% wsimport -keep -p  ch02.tc tempc.wsdl
```

generates the *wsimport* artifacts in the subdirectory *ch02/tc*. Of interest is the Java counterpart to the WSDL **portType**; in this case it is the file *ServiceSoap.java*, which contains the segment:

```
@WebService(name = "ServiceSoap",
            targetNamespace = "http://tempConvertURI.org/")
public interface ServiceSoap {
    @WebMethod(operationName = "c2f",
               action = "http://tempConvertURI.org/c2f")
    @WebResult(name = "c2fResult",
               targetNamespace = "http://tempConvertURI.org/")
    @RequestWrapper(localName = "c2f",
                    targetNamespace = "http://tempConvertURI.org/",
                    className = "ch02.tc.C2F")
    @ResponseWrapper(localName = "c2fResponse",
                     targetNamespace = "http://tempConvertURI.org/",
                     className = "ch02.tc.C2FResponse")
    public double c2F(
        @WebParam(name = "t", targetNamespace = "http://tempConvertURI.org/")
        double t);
    ...
```

This interface is the SEI, but it is easily transformed into the SIB as well: first, the keyword **interface** is changed to the keyword **class**; second, the method declarations

---

(only the declaration for c2F is shown here) are changed to method definitions. A third but optional step is to drop the service name ServiceSoap in favor of something more suggestive such as the name TempConvert. Here is the SIB that results from changes to the SEI:

```
package ch02.tc;

import javax.jws.WebMethod;
import javax.jws.WebParam;
import javax.jws.WebResult;
import javax.jws.WebService;
import javax.xml.ws.RequestWrapper;
import javax.xml.ws.ResponseWrapper;
@WebService(name = "TempConvert", targetNamespace = "http://tempConvertURI.org/")
public class TempConvert {
    @WebMethod(operationName = "c2f",
               action = "http://tempConvertURI.org/c2f")
    @WebResult(name = "c2fResult",
               targetNamespace = "http://tempConvertURI.org/")
    @RequestWrapper(localName = "c2f",
                    targetNamespace = "http://tempConvertURI.org/",
                    className = "ch02.tc.C2F")
    @ResponseWrapper(localName = "c2fResponse",
                     targetNamespace = "http://tempConvertURI.org/",
                     className = "ch02.tc.C2FResponse")
    public double c2F(
        @WebParam(name = "t",
                  targetNamespace = "http://tempConvertURI.org/")
        double t) { return 32.0 + (t * 9.0 / 5.0); }

    @WebMethod(operationName = "f2c",
               action = "http://tempConvertURI.org/f2c")
    @WebResult(name = "f2cResult",
               targetNamespace = "http://tempConvertURI.org/")
    @RequestWrapper(localName = "f2c",
                    targetNamespace = "http://tempConvertURI.org/",
                    className = "ch02.tc.F2C")
    @ResponseWrapper(localName = "f2cResponse",
                     targetNamespace = "http://tempConvertURI.org/",
                     className = "ch02.tc.F2CResponse")
    public double f2C(
        @WebParam(name = "t",
                  targetNamespace = "http://tempConvertURI.org/")
        double t) { return (5.0 / 9.0) * (t - 32.0); }

}
```

Once this Java-based web service is published in the usual way using the method Endpoint.publish, the *wsimport* utility can be used again but this time to generate client-side support code:

```
% wsimport -keep -p clientTC http://localhost:5599/tc?wsdl
```

A Java client such as ClientTC below then can be coded:

```
import javax.xml.ws.Service;
import clientTC.TempConvertService;
import clientTC.TempConvert;

class ClientTC {
    public static void main(String args[ ]) throws Exception {
        TempConvertService service = new TempConvertService();
        TempConvert port = service.getTempConvertPort();
        double d1 = -40.1, d2 = -39.4;
        System.out.printf("f2C(%f) = %f\n", d1, port.f2C(d1));
        System.out.printf("c2F(%f) = %f\n", d2, port.c2F(d2));
    }
}
```

When executed, this client outputs:

```
f2C(-40.100000) = -40.055556
c2F(-39.400000) = -38.920000
```

The example shows that the artifacts generated from *wsimport* can be used to support a Java client against a web service or to implement web service in Java. With a WSDL in hand, a Java implementation of the service is within reach.

A Code-First, Contract-Aware Approach

JWS also supports a *code-first, contract-aware* approach. JWS encourages the *code first* by making it easy to generate the WSDL. Once the service is published, the WSDL is generated automatically and available to clients. Java, however, does provide annotations that the programmer can use in order to determine, in critical areas, how the generated WSDL or, in turn, WSDL-generated artifacts will turn out. Example 2-23 shows yet another revision of the TimeServer service. For illustration, the web service is now implemented as a single file, *TimeServer.java*, and various annotations have been added so that their impact on the automatically generated WSDL and on SOAP messages can be seen.

Example 2-23. A code-first, contract-aware service

```
package ch02.tsa;  // 'a' for 'annotation'

import java.util.Date;
import javax.jws.WebService;
import javax.jws.WebMethod;
import javax.jws.Oneway;
import javax.jws.WebParam;
import javax.jws.WebParam.Mode;
import javax.jws.WebResult;
import javax.jws.soap.SOAPBinding;
import javax.jws.soap.SOAPBinding.Style;
import javax.jws.soap.SOAPBinding.Use;
import javax.jws.soap.SOAPBinding.ParameterStyle;
```

```
@WebService(name           = "AnnotatedTimeServer",
            serviceName    = "RevisedTimeServer",
            targetNamespace = "http://ch02.tsa")
@SOAPBinding(style          = SOAPBinding.Style.DOCUMENT,
             use            = SOAPBinding.Use.LITERAL,
             parameterStyle = SOAPBinding.ParameterStyle.WRAPPED)
public class TimeServer {
    @WebMethod(operationName   = "time_string")
    @WebResult(name            = "ts_out",
               targetNamespace = "http://ch02.tsa")
    public String getTimeAsString(
        @WebParam(name            = "client_message",
                  targetNamespace = "http://ch02.tsa",
                  mode            = WebParam.Mode.IN)
        String msg) {
            return msg + " at " + new Date().toString();
    }

    @WebMethod (operationName = "time_elapsed")
    public long getTimeAsElapsed() { return new Date().getTime(); }

    @WebMethod
    @Oneway
    public void acceptInput(String msg) { System.out.println(msg); }
}
```

Let's begin with the @WebService annotation, which includes attribute settings in this revision. Once the publisher application is running, with the revised TimeServer service available at port 8888, the command:

```
% wsimport -keep -p clientA http://localhost:8888/tsa?wsdl
```

generates the usual artifacts in subdirectory *clientA*. In the @WebService annotation, the:

```
serviceName = RevisedTimeServer
```

attribute causes the service artifact to be the class named RevisedTimeServer, and the:

```
name = AnnotatedTimeServer
```

attribute causes the portType artifact to be the class named AnnotatedTimeServer. The Java client against the revised service illustrates these points:

```
import clientA.RevisedTimeServer;
import clientA.AnnotatedTimeServer;

class TimeClientA {
    public static void main(String[ ] args) {
        RevisedTimeServer ts = new RevisedTimeServer();
        AnnotatedTimeServer ats = ts.getAnnotatedTimeServerPort();

        System.out.println(ats.timeString("Hi, world!"));
        System.out.println(ats.timeElapsed());
        ats.acceptInput("Hello, world!");
    }
}
```

In the revised `TimeServer`, the `targetNamespace` is set to `http://ch02.tsa` (with no trailing slash) ensures that this will be the namespace URI for the service in the WSDL; in other words, the `ch02.tsa` will not be inverted to `tsa.ch02`, which occurred in effect in the original `TimeServer` service. The `@SOAPBinding` attributes, set to their default values, are included for illustration.

The `getTimeAsString` method is the most heavily annotated of the three `@WebMethod`s, with both a `@WebResult` and a `@WebParam` annotation. This method and the method `getTimeAsElapsed` have their operational names explicitly set to `time_string` and `time_elapsed`, respectively. Here is the segment from the WSDL's `message` section that reflects these settings. Note, too, that the message names reflect the operation names, as is standard in the wrapped `document` style:

```
<message name="time_string">
  <part element="tns:time_string" name="parameters"></part>
</message>
<message name="time_stringResponse">
  <part element="tns:time_stringResponse" name="parameters"></part>
</message>
<message name="time_elapsed">
  <part element="tns:time_elapsed" name="parameters"></part>
</message>
<message name="time_elapsedResponse">
  <part element="tns:time_elapsedResponse" name="parameters"></part>
</message>
```

The `@WebResult` annotation of the `getTimeAsString` method does not impact the WSDL but rather the SOAP response message; in particular the subelement with the tag name `ns1:ts_out` in the wrapped SOAP body:

```
<?xml version="1.0" ?>
<soapenv:Envelope xmlns:soapenv="http://schemas.xmlsoap.org/soap/envelope/"
                  xmlns:xsd="http://www.w3.org/2001/XMLSchema"
                  xmlns:ns1="http://ch02.tsa">
  <soapenv:Body>
    <ns1:time_stringResponse>
      <ns1:ts_out>Hi, world! at Thu Oct 23 22:24:59 CDT 2008</ns1:ts_out>
    </ns1:time_stringResponse>
  </soapenv:Body>
</soapenv:Envelope>
```

The `@WebMethod` named `acceptInput` is annotated as `@Oneway`, which requires that the method return `void`. As a result, a client can send a request to the parameterized `acceptInput` operation but receives no response. Following is the WSDL segment that defines the operation with only an `input` message and no `output` message.

```
<operation name="acceptInput">
  <input message="tns:acceptInput"></input>
</operation>
```

The example shows some of the possibilities in JWS's *code-first, contract-aware* approach. If the web service creator is willing to accept defaults, the JWS annotations such as `@WebService` and `@WebMethod` are clear and simple. If fine-grained control over the WSDL and the SOAP messages is required, such control is available with additional annotations.

Limitations of the WSDL

WSDL documents, as web service descriptions, should be publishable and discoverable. A *UDDI* (Universal Description Discovery and Integration) registry is one way to publish WSDLs so that potential clients can discover them and ultimately consume the services that the WSDLs describe. UDDI does not directly support any particular type of service description, including WSDL, but instead provides its own type system that accommodates WSDL documents. In UDDI terms, a WSDL is essentially a two-part document. The first part, which comprises the `types` through the `binding` sections, is the UDDI *service interface*. The second part, which comprises any `import` directives and the `service` section, is the UDDI *service implementation*. In WSDL, the service interface and service implementation are two parts of the same document. In UDDI, the two parts are separate documents.

Once a WSDL has been located, perhaps through a UDDI service, critical questions remain about the service described in the WSDL. For one, the WSDL does not explain service *semantics* or, in plainer terms, what the service is about. The WSDL precisely describes what might be called the service's invocation syntax: the names of the service operations; the expected pattern for a service operation (e.g., request/response); the number, order, and type of arguments that each operation expects; fault codes, if any, associated with a service operation; and the number and types of any response values from the service. The Amazon E-Commerce WSDL, which has about 3,640 lines, contains such information for all of the many operations that the service provides. Yet the WSDL itself contains no information about the intended use of the service. Figuring this out is left to the programmer, who, drawing presumably on experience with the Amazon website, recognizes from service operation names such as `itemSearch`, `sellerLookup`, and `cartCreate` that the E-Commerce service is meant to replicate the functionality available from a browser-based session at the Amazon website. Amazon does provide supplementary material such as documentation, tutorials, and sample code libraries, all of which are meant to provide the semantic information that the WSDL itself does not provide. The W3C is pursuing initiatives in web semantics under the rubric of WSDL-S (Semantics). For more information on WSDL-S, see *http://www .w3.org/Submission/WSDL-S*. As of now, a WSDL typically is useful only if the client programmer already understands what the service is about.

What's Next?

SOAP-based web services in Java have two levels. The application level, which consists of the service itself and any Java-based client, typically and appropriately hides the SOAP. The handler level, which consists of SOAP message interceptors on either the service or the client side, can manipulate the SOAP. At times it is useful to handle the SOAP directly, as the next chapter illustrates. Chapter 3 also looks at how SOAP-based web services can transport large binary payloads efficiently.

## CHAPTER 3

SOAP Handling

SOAP: Hidden or Not?

Until now we have looked at SOAP messages only to see how SOAP-based web services operate under the hood. After all, the point of JWS is to develop and consume web services, ignoring infrastructure whenever possible. It is one thing to track web service messages at the wire level in order to see what is really going on. It is quite another to process SOAP messages in support of application logic.

At times, however, a web service or a client might need to process a SOAP envelope. For example, a client might need to add security credentials to the header of a SOAP message before sending the message; and the web service that receives the message then might need to validate these security credentials after extracting them from the header of the incoming SOAP message. How to add, extract, and otherwise process information in a SOAP message is the main topic in this chapter. The goal is to illustrate techniques for dealing directly with SOAP messages in SOAP-based web services. This chapter also examines SOAP attachments, particularly for large binary payloads.

SOAP 1.1 and SOAP 1.2

SOAP now comes in two versions, 1.1 and 1.2. There is debate within the W3C about whether a version 1.3 is needed. The good news is that the differences between SOAP 1.1 and SOAP 1.2 have minor impact on programming web services in general and programming JAX-WS in particular. There are exceptions. For example, the structure of the SOAP header differs between the versions, which can impact programming. Figure 3-1 depicts the general structure of a SOAP message under either 1.1 or 1.2.

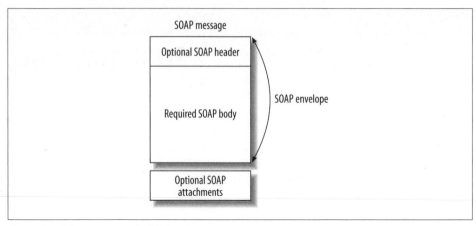

Figure 3-1. The structure of a SOAP Message

Under either version, SOAP messages can be manipulated through *handlers*, which are programmer-written classes that contain *callbacks*; that is, methods invoked from the web service runtime so that an application has access to the underlying SOAP. At what might be called the application level, which consists of web services and their clients, the SOAP remains hidden. At the handler level, the SOAP is exposed so that the programmer can manipulate the incoming and outgoing messages.

Even at the handler level, most differences between 1.1 and 1.2 amount to refinements that usually can be ignored. For example, SOAP 1.1 technically allows XML elements to occur *after* the SOAP body. In SOAP 1.2, by contrast, the SOAP body is the last XML element in the SOAP envelope. Even in SOAP 1.1, however, it is only in contrived examples that a SOAP envelope contains XML elements in the wasteland between the end of the SOAP body and the end of the SOAP envelope. The 1.1 version binds SOAP to HTTP transport, whereas the 1.2 version also supports SOAP over SMTP. The SOAP 1.1 specification is a single document, but the SOAP 1.2 specification is divided into three documents. In JAX-WS, as in most frameworks, SOAP 1.1 is the default and SOAP 1.2 is an option readily available, as examples in this chapter illustrate. Finally, starting with the 1.2 version, *SOAP* is officially no longer an acronym!

SOAP Messaging Architecture

A SOAP message is a one-way transmission from a sender to a receiver; hence, the fundamental message exchange pattern (MEP) for SOAP is one way. SOAP-based applications such as web services are free to set up conversational patterns that combine one-way messaging in richer ways. The request/response MEP in a SOAP-based web service is a brief conversation in which a request initiates the conversion and a response concludes the conversation. MEPs such as request/response and solicit/response can be put together in suitable ways to support more expansive conversational patterns as needed.

Although a SOAP message is intended for an ultimate receiver, the SOAP messaging architecture allows for *SOAP intermediaries*, which are nonterminal recipients or *nodes* along the route from the sender to the ultimate receiver. An intermediary may inspect and even manipulate an incoming SOAP message before sending the message on its way toward the ultimate receiver. Figure 3-2 depicts a SOAP sender, two intermediaries, and an ultimate receiver.

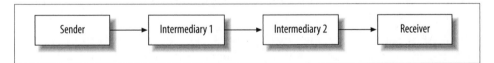

Figure 3-2. Sender, intermediaries, and receiver

Recall that a SOAP envelope has a required body, which may be empty, and an optional header. An intermediary should inspect and process only the elements in the SOAP header rather than anything in the SOAP body, which carries whatever cargo the sender intends for the ultimate receiver alone. The header, by contrast, is meant to carry whatever meta-information is appropriate for either the ultimate receiver or intermediaries. For example, the header might contain the sender's digital signature as a voucher or include a timestamp that indicates when the information in the message's body becomes obsolete. XML elements within the optional header are *header blocks* in SOAP speak. Example 3-1 shows a SOAP message with a SOAP 1.1 header, which in turn contains a header block tagged `uuid`.

Example 3-1. A sample SOAP header

```
<S:Envelope xmlns:S="http://schemas.xmlsoap.org/soap/envelope/">
   <S:Header>
      <uuid xmlns="http://ch03.fib"
            xmlns:SOAP-ENV="http://schemas.xmlsoap.org/soap/envelope/"
            SOAP-ENV:actor="http://schemas.xmlsoap.org/soap/actor/next">
         ca12fd33-16e1-4a95-b17e-3ef6744babdc
      </uuid>
   </S:Header>
   <S:Body>
      <ns2:countRabbits xmlns:ns2="http://ch03.fib">
         <arg0>45</arg0>
      </ns2:countRabbits>
   </S:Body>
</S:Envelope>
```

The header block tagged with `uuid` contains a *UUID* (Universally Unique IDentifier) value, which is a 128-bit number formatted as a string with hexadecimal numerals. As the name suggests, a UUID is meant to be a statistically unique identifier, in this case an identifier for this particular request against the `RabbitCounter` web service, which is introduced shortly. The assumption in this example is that every request against the `RabbitCounter` service would have its own UUID, which might be logged with other information for later analysis.

The header block with the `uddi` tag has an attribute, `SOAP-ENV:actor`, whose value ends with `next`. The `next` actor, in this scheme, is the next recipient on the path from the message sender to the ultimate receiver; hence, each intermediary (and the ultimate receiver) acts in a `next` role. A `next` actor is expected to inspect header blocks in some application-appropriate way. If a node cannot process the header block, then the node should throw a fault. In SOAP 1.2, this expectation can be made explicit by setting the `mustUnderstand` attribute of a header block to `true`, which means that a node *must* throw a fault if the node cannot process the header block.

The SOAP specification does not explain exactly how an intermediate or final node must process a header block, as this is an application-specific rather than a SOAP requirement. In this example, the next node might check whether the UUID value is well-formed under an application-mandated algorithm such as *SHA-1* (Secure Hash Algorithm-1). If the value cannot be verified as well-formed, then the node should throw a fault to signal this fact.

Of interest here is how a header such as the one in Example 3-1 can be generated. By default, JWS generates a SOAP request message without a header. The next section examines how application code can generate the SOAP message in Example 3-1.

Programming in the JWS Handler Framework

JWS provides a *handler framework* that allows application code to inspect and manipulate outgoing and incoming SOAP messages. A handler can be injected into the framework in two steps:

1. One step is to create a handler class, which implements the `Handler` interface in the `javax.xml.ws.handler` package. JWS provides two `Handler` subinterfaces, `LogicalHandler` and `SOAPHandler`. As the names suggest, the `LogicalHandler` is protocol-neutral but the `SOAPHandler` is SOAP-specific. A `LogicalHandler` has access only to the message payload in the SOAP body, whereas a `SOAPHandler` has access to the entire SOAP message, including any optional headers and attachments. The class that implements either the `LogicalHandler` or the `SOAPHandler` interface needs to define three methods for either interface type, including `handleMessage`, which gives the programmer access to the underlying message. The other two shared methods are `handleFault` and `close`. The `SOAPHandler` interface requires the implementation to define a fourth method, `getHeaders`. These methods are explained best through code examples.

2. The other step is to place a handler within a handler chain. This is typically done through a configuration file, although handlers also can be managed through code. An example follows shortly.

Once injected into the handler framework, a programmer-written handler acts as a message interceptor that has access to every incoming and outgoing message. Under the request/response message exchange pattern (MEP), for instance, a client-side handler has access to the outgoing request message after the message has been created but before the message is sent to the web service. The same client-side handler has access to the incoming response message from the web service. A service-side handler under this MEP has access to the incoming request message and to the outgoing response message after the response message has been created.

The JWS handler framework thus encourages the *chain of responsibility* pattern, which Java servlet programmers encounter when using filters. The underlying idea is to distribute responsibility among various handlers so that the overall application is highly modular and, therefore, more easily maintainable.

The RabbitCounter Example

Now we can drill down into the details with an example, which introduces constructs that will be useful in later examples. Following is a summary of what the example does:

- The `RabbitCounter` service has one operation, `countRabbits`, that expects a single integer argument and returns an integer. The operation computes the Fibonacci numbers, which are relevant in botany, engineering, computer science, aesthetics, market trading, and even rabbit breeding. The sample operation needs to be parameterized so that SOAP faults can be illustrated. In this example, a SOAP fault is thrown if the argument is a negative integer. The example thus provides a first look at SOAP faults.

- The service, its clients, and any intermediaries along the route from client to service are expected to process SOAP headers. In particular, a client injects a header block into an outgoing (that is, request) message and any intermediary and the ultimate receiver validate the information in the header block, generating a SOAP fault if necessary. In later examples, header blocks will be used to carry credentials such as digital signatures. For now, the details of injecting and processing header blocks are of primary interest.

- JWS has two different ways to throw SOAP faults and the example illustrates both. The simplest way is to extend the `Exception` class (for example, with a class named `FibException`) and to throw the exception in a `@WebMethod` whenever appropriate. JWS then automatically maps the Java exception to a SOAP fault. The other way, which takes more work, is to throw a fault from a handler. In this case, a `SOAPFaultException` is created and then thrown.

The following sections flesh out the details.

Injecting a Header Block into a SOAP Header

The header block in Example 3-1 comes from the client-side handler shown in Example 3-2.

Example 3-2. A handler that injects a SOAP header block

```
package fibC;

import java.util.UUID;
import java.util.Set;
import java.util.logging.Logger;
import javax.xml.namespace.QName;
import javax.xml.soap.SOAPMessage;
import javax.xml.ws.handler.MessageContext;
import javax.xml.ws.handler.soap.SOAPHandler;
import javax.xml.ws.handler.soap.SOAPMessageContext;
import javax.xml.soap.SOAPEnvelope;
import javax.xml.soap.SOAPHeader;
import javax.xml.soap.SOAPHeaderElement;
import javax.xml.soap.SOAPException;
import javax.xml.soap.SOAPConstants;
import java.io.IOException;

public class UUIDHandler implements SOAPHandler<SOAPMessageContext> {
    private static final String LoggerName = "ClientSideLogger";
    private Logger logger;
    private final boolean log_p = true; // set to false to turn off
    public UUIDHandler() {
        logger = Logger.getLogger(LoggerName);
    }

    public boolean handleMessage(SOAPMessageContext ctx) {
        if (log_p) logger.info("handleMessage");

        // Is this an outbound message, i.e., a request?
        Boolean request_p = (Boolean)
          ctx.get(MessageContext.MESSAGE_OUTBOUND_PROPERTY);

        // Manipulate the SOAP only if it's a request
        if (request_p) {
            // Generate a UUID and a timestamp to place in the message header.
            UUID uuid = UUID.randomUUID();

            try {
                SOAPMessage msg = ctx.getMessage();
                SOAPEnvelope env = msg.getSOAPPart().getEnvelope();
                SOAPHeader hdr = env.getHeader();
                // Ensure that the SOAP message has a header.
                if (hdr == null) hdr = env.addHeader();

                QName qname = new QName("http://ch03.fib", "uuid");
                SOAPHeaderElement helem = hdr.addHeaderElement(qname);

                helem.setActor(SOAPConstants.URI_SOAP_ACTOR_NEXT); // default
                helem.addTextNode(uuid.toString());
                msg.saveChanges();

                // For tracking, write to standard output.
                msg.writeTo(System.out);
            }
```

```
            catch(SOAPException e) { System.err.println(e); }
            catch(IOException e) { System.err.println(e); }
        }
         return true; // continue down the chain
    }

    public boolean handleFault(SOAPMessageContext ctx) {
        if (log_p) logger.info("handleFault");
        try {
            ctx.getMessage().writeTo(System.out);
        }
        catch(SOAPException e) { System.err.println(e); }
        catch(IOException e) { System.err.println(e); }
        return true;
    }

    public Set<QName> getHeaders() {
        if (log_p) logger.info("getHeaders");
        return null;
    }

    public void close(MessageContext messageContext) {
        if (log_p) logger.info("close");
    }
}
```

Because the UUIDHandler is a client-side handler and the MEP is request/response, the handleMessage method is invoked by the framework *after* the underlying SOAP request message has been constructed but *before* the request message has been sent to the service. Suppose that this handler is configured to work with TestClient, which is the client of a SOAP-based service TestService. The details of the configuration will be given shortly. For now, here is a summary of what happens given a TestClient against a TestService and a client-side UUIDHandler:

1. Whenever the TestClient generates a request against the TestService, the client-side JWS libraries create a SOAP message that serves as the request.

2. Once the SOAP message has been created but before the message is sent, the UUIDHandler callbacks are invoked. In particular, the callback handleMessage has access to the full SOAP message because the encapsulating class UUIDHandler is a SOAPHandler rather than a LogicalHandler, which would have access only to the message payload. The handleMessage callback injects a UUID value into the header of the SOAP message.

3. After the UUIDHandler has done its work, the SOAP request message with the injected header block is sent on its way to the ultimate receiver, which in this case is the TestService.

Here are some additional details about the process. The same UUIDHandler has access to the incoming or response message. The method handleMessage has an argument SOAPMessageContext, which gives access to the underlying SOAP message. Accordingly, the `handleMessage` callback first checks whether the SOAP is outgoing (that is, a request from the client's perspective). For an outgoing message, the handler generates a UUID value to place in a header block. One problem is that the SOAP header is *optional* under either SOAP 1.1 or SOAP 1.2. So the handler checks whether the already-created SOAP message has a header and, if not, adds a header to the SOAP envelope. Next, an instance of a `SOAPHeaderElement`, the Java implementation of a header block, is added to the SOAP header, and the `actor` attribute is added using a `set` method. Here is the code segment:

```
SOAPHeaderElement helem = hdr.addHeaderElement(qname);
helem.setActor(SOAPConstants.URI_SOAP_ACTOR_NEXT); // the default
```

The UUID value is added to the header as an XML text node and the overall changes to the SOAP message are saved. The outgoing SOAP message reflects the changes. Here, for review, is the relevant segment of the SOAP header after the handler executes and the additions have been saved:

```
<S:Header>
    <uuid xmlns="http://ch03.fib"
          xmlns:SOAP-ENV="http://schemas.xmlsoap.org/soap/envelope/"
          SOAP-ENV:actor="http://schemas.xmlsoap.org/soap/actor/next">
        ca12fd33-16e1-4a95-b17e-3ef6744babdc
    </uuid>
</S:Header>
```

For now, the `UUIDHandler` in Example 3-2 happens to be the only handler in the *handler chain* (that is, the sequence of handlers) active in the service client, `FibClient`. There could be other handlers, each with a position within the chain. Although the handler chain can be set up in code, it is cleaner to describe the chain in a deployment file. Example 3-3 is the deployment file *handler-chain.xml*, although any name will do.

Example 3-3. A handler deployment file

```
<?xml version="1.0" encoding="UTF-8" standalone="yes"?>
<javaee:handler-chains
     xmlns:javaee="http://java.sun.com/xml/ns/javaee"
     xmlns:xsd="http://www.w3.org/2001/XMLSchema">
  <javaee:handler-chain>
    <javaee:handler>
       <javaee:handler-class>fibC.UUIDHandler</javaee:handler-class>
    </javaee:handler>
  </javaee:handler-chain>
</javaee:handler-chains>
```

If there were other handlers, any mix of `LogicalHandler` and `SOAPHandler` implementations, these would be listed before or after the listing of `fibC.UUIDHandler`, the fully qualified name of the class in Example 3-2. Handler methods of one type (for instance, `handleMessage` methods in `SOAPHandlers`) execute in order, from top to bottom, unless one of the executing handler methods halts handler processing altogether. A handler method halts execution of the remaining handlers in the chain as follows. The methods `handleMessage` and `handleFault` return `boolean` values. A returned `true` means *continue message processing by executing the next handler in the chain*, whereas a returned `false` means *stop message processing*. If a handler stops processing on an outgoing message (for instance, a client request message or a service response message), then the message is not sent.

The top-to-bottom sequence of the handlers in the configuration file determines the order in which handler methods of one type (e.g., `SOAPHandler`) execute. The runtime ordering may differ from the order given in the configuration file. Here is the reason:

- For an outbound message (for instance, a client request under the request/response MEP), the `handleMessage` method or `handleFault` method in a `LogicalHandler` code execute *before* their counterparts in a `SOAPHandler`.

- For an inbound message, the `handleMessage` method or `handleFault` method in a `SOAPHandler` code execute *before* their counterparts in a `LogicalHandler`.

For example, suppose that a handler configuration file lists SOAP handlers (SH) and logical handlers (LH) in this order, top to bottom:

```
SH1
LH1
SH2
SH3
LH2
```

Despite the order in the configuration file, the handlers execute in this order, top to bottom, on an *outgoing* message:

```
LH1
LH2
SH1
SH2
SH3
```

On an *incoming* message, the handlers execute in this order, top to bottom:

```
SH1
SH2
SH3
LH1
LH2
```

Figure 3-3 depicts the runtime ordering of logical and SOAP handlers. This runtime ordering makes sense because the `LogicalHandler` has access only to the body of the SOAP message, whereas the `SOAPHandler` has access to the entire SOAP message. For an outgoing message, then, the logical handlers should be able to process the payload, the SOAP body, before the SOAP handlers process the entire SOAP message. If the application does not need to process SOAP headers or SOAP attachments, then a `LogicalHandler` is the way to go.

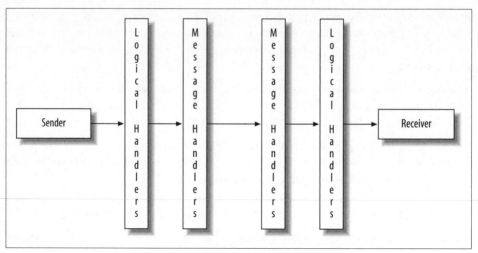

Figure 3-3. Organization of logical and message handlers

The deployment file, in this case *handler-chain.xml*, can be located anywhere on the classpath. In this example, the file is in the *fibC* subdirectory, which is the named package for the *wsimport*-generated stubs. The compiled client, *FibClient.class*, is in the parent directory of *fibC*. The reasons for this layout are given in the next section. In any case, here is the FibClient code:

```
import fibC.RabbitCounterService;
import fibC.RabbitCounter;

class FibClient {
    public static void main(String[ ] args) {
        RabbitCounterService service = new RabbitCounterService();
        RabbitCounter port = service.getRabbitCounterPort();
        try {
            int n = -45;
            System.out.println("fib(" + n + ") = " + port.countRabbits(n));
        }
        catch(Exception e) { System.err.println(e); }
    }
}
```

The call to the web service method countRabbits occurs in a try block because this method throws a SOAP fault on a negative argument, as in this case. Otherwise, the client pattern is quite familiar by now.

Finally, the UUIDHandler also is used to provide a wire-level dump of the outgoing SOAP message. In Java, SOAP handlers are thus an alternative to a dump utility such as *tcpmon* or *tcpdump*.

Configuring the Client-Side SOAP Handler

The `FibClient` sends requests to the `ch03.fib.RabbitCounter` web service, as shown in Example 3-4.

Example 3-4. The RabbitCounter service

```java
package ch03.fib;

import java.util.Map;
import java.util.HashMap;
import java.util.Collections;
import javax.jws.WebService;
import javax.jws.WebMethod;
import javax.jws.soap.SOAPBinding;
import javax.jws.soap.SOAPBinding.Style;
import javax.jws.soap.SOAPBinding.Use;
import javax.jws.soap.SOAPBinding.ParameterStyle;

@WebService(targetNamespace = "http://ch03.fib")
@SOAPBinding(style          = SOAPBinding.Style.DOCUMENT,
             use            = SOAPBinding.Use.LITERAL,
             parameterStyle = SOAPBinding.ParameterStyle.WRAPPED)
public class RabbitCounter {
    // stores previously computed values
    private Map<Integer, Integer> cache =
        Collections.synchronizedMap(new HashMap<Integer, Integer>());

    @WebMethod
    public int countRabbits(int n) throws FibException {
        // Throw a fault if n is negative.
        if (n < 0) throw new FibException("Neg. arg. not allowed.", n + " < 0");

        // Easy cases.
        if (n < 2) return n;

        // Return cached values if present.
        if (cache.containsKey(n)) return cache.get(n);
        if (cache.containsKey(n - 1) &&
            cache.containsKey(n - 2)) {
          cache.put(n, cache.get(n - 1) + cache.get(n - 2));
          return cache.get(n);
        }

        // Otherwise, compute from scratch, cache, and return.
        int fib = 1, prev = 0;
        for (int i = 2; i <= n; i++) {
            int temp = fib;
            fib += prev;
            prev = temp;
        }
        cache.put(n, fib); // cache value for later lookup
        return fib;
    }
}
```

The service has only one operation, `countRabbits`. The Fibonacci numbers can be computed using recursion, but this technique is inefficient. Although this recursive definition of the Fibonacci numbers:

```
          undefined if n < 0
fib(n) = n if 0 <= n < 2
          fib(n - 1) + fib(n - 2) otherwise
```

converts straightforwardly into a Java method:

```
int fib(int n) {
    if (n < 0) throw new RuntimeException("Undefined for negative values.");
    if (n < 2) return n;                // base case
    else return fib(n - 1) + fib(n - 2);  // recursive calls
}
```

the recursive implementation repeats computations. For instance, the call to `fib(5)` computes `fib(2)` three separate times under the recursive implementation. The *base case* for the recursive definition of the `fib` function occurs when n = 1. If n >= 2, then `fib(n)` computes the base case `fib(n - 1)` times, which is highly inefficient for large values of n. Although the `RabbitCounter` service may be whimsical, there are common-sensical applications of the Fibonacci numbers. Suppose, for instance, that someone is able to take a normal step of 1 meter and a jump step of 2 meters. How many ways can this person traverse, say, a 100 meters? The answer is `fib(100)`: 3,314,859,971.

The web service operation `countRabbits` does not recompute any values in computing `fib(n)`. Further, this implementation caches previously computed values so that these may be looked up rather than computed from scratch in subsequent calls.

Our real interest is in the client-side handlers, which intercept the request SOAP messages sent to the `RabbitCounterService` and the response SOAP messages returned from this service. The next step is to indicate where the configuration file for the handler chain is to be found. The *wsimport*-generated stub for the service, `fibC.RabbitCounterService`, is an obvious place to point to the configuration file. Here is the `@HandlerChain` annotation added to the stub:

```
import javax.jws.HandlerChain;
@WebServiceClient(name = "RabbitCounterService",
                  targetNamespace = "http://ch03.fib",
                  wsdlLocation = "http://localhost:8888/fib?wsdl")
@HandlerChain(file = "handler-chain.xml")
public class RabbitCounterService extends Service {
```

The configuration file can be anywhere on the classpath; hence, it is convenient to place it in the same *fibC* subdirectory that holds the *wsimport*-generated files.

Adding a Handler Programmatically on the Client Side

Managing handlers through a configuration file is the preferred way but not the only way. It is preferred because this way keeps the client or service code relatively clean. Handlers are at the edge of application logic rather than at the center; hence, they are usually managed best through metadata files such as *handler-chain.xml* in the current example. Nonetheless, it is not hard to manage handlers programmatically.

Example 3-5 is a revised client against the `RabbitCounter` service.

Example 3-5. A client that configures a handler programmatically

```java
import fibC.RabbitCounterService;
import fibC.RabbitCounter;
import fibC.UUIDHandler;
import fibC.TestHandler;

import java.util.List;
import java.util.ArrayList;
import javax.xml.ws.handler.Handler;
import javax.xml.ws.handler.HandlerResolver;
import javax.xml.ws.handler.PortInfo;

class FibClientHR {
    public static void main(String[ ] args) {
        RabbitCounterService service = new RabbitCounterService();
        service.setHandlerResolver(new ClientHandlerResolver());
        RabbitCounter port = service.getRabbitCounterPort();

        try {
            int n = 27;
            System.out.printf("fib(%d) = %d\n", n, port.countRabbits(n));
        }
        catch(Exception e) { System.err.println(e); }
    }
}
class ClientHandlerResolver implements HandlerResolver {
    public List<Handler> getHandlerChain(PortInfo port_info) {
        List<Handler> hchain = new ArrayList<Handler>();
        hchain.add(new UUIDHandler());
        hchain.add(new TestHandler()); // for illustration only
        return hchain;
    }
}
```

The file *FibClientHR.java* has two classes, `FibClientHR` and `ClientHandlerResolver`. The `HandlerResolver` interface declares only one method, `getHandlerChain`, which the runtime invokes to get a list of the handlers. In `FibClientHR`, the `setHandlerResolver` method is invoked on the `service` object:

```java
service.setHandlerResolver(new ClientHandlerResolver());
```

As a result, the configuration file *handle-chain.xml* now plays no role in the handler configuration. Instead, the code takes on this management role. The example adds a second handler, `TestHandler`, to illustrate the intuitive ordering. The `UUIDHandler` comes first in the handler chain and so is executed first. By the way, the `TestHandler` simply prints the SOAP message to the standard output for tracking purposes.

Generating a Fault from a @WebMethod

The RabbitCounter service includes a customized exception that the @WebMethod named
countRabbits throws if the method is invoked with a negative integer as an argument.
Here is the code segment in the method definition:

```
@WebMethod
public int countRabbits(int n) throws FibException {
   // Throw a fault if n is negative.
   if (n < 0) throw new FibException("Negative args not allowed.", n + " < 0");
```

And here is the customized exception class:

```
package ch03.fib;

public class FibException extends Exception {
    private String details;
    public FibException(String reason, String details) {
      super(reason);
      this.details = details;
    }
    public String getFaultInfo() { return details; }
}
```

If a FibException is thrown in the countRabbits operation, a fault message rather than
an output message becomes the service's response to the client. For example, here is
the response from a call with -999 as the argument:

```
<S:Envelope xmlns:S="http://www.w3.org/2003/05/soap-envelope">
  <S:Header/>
  <S:Body>
    <ns3:Fault xmlns:ns2="http://schemas.xmlsoap.org/soap/envelope/"
               xmlns:ns3="http://www.w3.org/2003/05/soap-envelope">
      <ns3:Code><ns3:Value>ns3:Receiver</ns3:Value></ns3:Code>
      <ns3:Reason>
        <ns3:Text xml:lang="en">Negative args not allowed.</ns3:Text>
      </ns3:Reason>
      <ns3:Detail>
        <ns2:FibException xmlns:ns2="http://ch03.fib">
          <faultInfo>-999 < 0</faultInfo>
          <message>Negative args not allowed.</message>
        </ns2:FibException>
      </ns3:Detail>
      ...
    </ns3:Fault>
  </S:Body>
</S:Envelope>
```

The SOAP fault message includes the Reason, which is the first argument in the constructor call:

```
new FibException("Negative args not allowed.", n + " < 0");
```

and additional `Detail`, which is the second argument in the constructor call. The response fault message is large, with a wealth of detail. The segment shown above contains the essentials.

When the service is published, the generated WSDL likewise reflects that the service can throw a SOAP fault. The WSDL includes a message to implement the exception and the `portType` section includes a `fault` message together with the usual `input` and `output` messages. Here are the relevant WSDL segments:

```
<message name="FibException">
   <part name="fault" element="tns:FibException"/>
</message>
<portType name="RabbitCounter">
   <operation name="countRabbits">
     <input message="tns:countRabbits"/>
     <output message="tns:countRabbitsResponse"/>
     <fault message="tns:FibException" name="FibException"/>
   </operation>
</portType>
```

Throwing a SOAP fault from a `@WebMethod` is straightforward. The `Exception`-based class that implements the SOAP fault in Java should have a constructor with two arguments, the first of which gives the reason for the fault (a negative argument, in this example) and the second of which provides additional details about the fault. The `Exception`-based class should define a `getFaultInfo` method that returns the fault details. The corresponding SOAP fault message includes the reason and the details.

Adding a Logical Handler for Client Robustness

The `RabbitCounter` service throws a SOAP fault if the method `countRabbits` is called with a negative argument. A `LogicalHandler` on the client side could intercept the outgoing request, check the argument to `countRabbits`, and change the argument if it is negative. The `ArgHandler` in Example 3-6 does just this.

Example 3-6. A logical handler for robustness

```
package fibC;

import javax.xml.ws.LogicalMessage;
import javax.xml.ws.handler.LogicalHandler;
import javax.xml.ws.handler.LogicalMessageContext;
import javax.xml.ws.handler.MessageContext;
import java.util.logging.Logger;

import javax.xml.bind.JAXBContext;
import javax.xml.bind.JAXBElement;
import javax.xml.bind.JAXBException;

public class ArgHandler implements LogicalHandler<LogicalMessageContext> {
    private static final String LoggerName = "ArgLogger";
    private Logger logger;
    private final boolean log_p = true; // set to false to turn off
```

```
    public ArgHandler() {
        logger = Logger.getLogger(LoggerName);
    }

    // If outgoing message argument is negative, make non-negative.
    public boolean handleMessage(LogicalMessageContext ctx) {
        Boolean outbound_p = (Boolean)
          ctx.get(MessageContext.MESSAGE_OUTBOUND_PROPERTY);
        if (outbound_p) {
            if (log_p) logger.info("ArgHandler.handleMessage");
            LogicalMessage msg = ctx.getMessage();

            try {
                JAXBContext jaxb_ctx = JAXBContext.newInstance("fibC");
                Object payload = msg.getPayload(jaxb_ctx);
                if (payload instanceof JAXBElement) {
                    Object obj = ((JAXBElement) payload).getValue();
                    CountRabbits obj_cr = (CountRabbits) obj;
                    int n = obj_cr.getArg0();       // current value
                    if (n < 0) {                    // negative argument?
                        obj_cr.setArg0(Math.abs(n)); // make non-negative

                        // Update the message.
                        ((JAXBElement) payload).setValue(obj_cr);
                        msg.setPayload(payload, jaxb_ctx);
                    }
                }
            }
            catch(JAXBException e) { }
        }
        return true;
    }
    public boolean handleFault(LogicalMessageContext ctx) { return true; }
    public void close(MessageContext ctx) { }
}
```

The `ArgHandler` uses a `JAXBContext` to extract the payload from the `LogicalHandler`. This
payload is the body of the outgoing SOAP message. There is also a no-argument version
of the `LogicalHandler` method `getPayload`, which returns an XML `Source` object. The
XML from this source could be unmarshaled into a Java object. For the one-argument
version of `getPayload` used here, the argument can be either a class or, as in this example,
the names of one or more packages. (Recall that `fibC` is the package name for the
wsimport-generated artifacts.) In any case, the message's payload is extracted as a
`JAXBElement`, which represents in XML a `CountRabbits` object, where `CountRabbits` is
one of the *wsimport*-generated artifacts that supports the client code. In particular, the
`fibC.CountRabbits` class is the Java type that corresponds to a SOAP request for
the web service operation exposed as `countRabbits`.

The `CountRabbits` class has two methods, `getArg0` and `setArg0`, that give access to the
argument passed to the `countRabbits` operation. If this argument is a negative integer,
then the logical handler produces the nonnegative value with a call to `Math.abs`, updates

the JAXBElement to reflect this change, and finally updates the message body with a call to setPayload. This handler makes clients more robust by intercepting SOAP requests that otherwise would cause the service to throw a SOAP fault.

The handler deployment file *handler-chain.xml* is updated to reflect a second handler in the chain. Here is the revised configuration file:

```
<?xml version="1.0" encoding="UTF-8" standalone="yes"?>
<javaee:handler-chains
     xmlns:javaee="http://java.sun.com/xml/ns/javaee"
     xmlns:xsd="http://www.w3.org/2001/XMLSchema">
  <javaee:handler-chain>
    <javaee:protocol-bindings>##SOAP12_HTTP</javaee:protocol-bindings>
    <javaee:handler>
      <javaee:handler-class>fibC.UUIDHandler</javaee:handler-class>
    </javaee:handler>
    <javaee:handler>
      <javaee:handler-class>fibC.ArgHandler</javaee:handler-class>
    </javaee:handler>
  </javaee:handler-chain>
</javaee:handler-chains>
```

The ArgHandler is listed deliberately *after* the UUIDHandler to underscore that, whatever the configuration sequence, the handleMessage or handleFault methods in a logical handler always execute *before* the corresponding methods in a SOAP handler on outgoing messages. For incoming messages, these SOAP handler methods execute before their logical handler counterparts.

The RabbitCounter service still checks for negative arguments and throws a SOAP fault if there is one, as the service obviously cannot count on the client-side logical handler. Of course, the web service itself might convert negative arguments to nonnegative ones, but the current service design lets us explore how JWS handlers can be put to good use on the client side.

Adding a Service-Side SOAP Handler

The client-side UUIDHandler inserts a header block with the actor attribute set to next. This setting means that any intermediary along the message path from sender to receiver should process the header block and, if appropriate, throw a fault. A fault is appropriate just in case the intermediary cannot process the header block in some way that the application requires. The SOAP specification does not explain the details of what processing an application may require of intermediaries and ultimate receivers; the specification merely allows senders to signal that application-specific processing of header blocks is expected along the path to the receiver. In this messaging architecture, the ultimate receiver also counts as a next actor and is likewise expected to process the header block and throw a fault, if needed. To complete the example, then, we need a service-side SOAP handler to process the header block that the client-side application inserts into the outgoing SOAP message. The service-side handler illustrates how an intermediary or the ultimate receiver might process the header block. The specific processing logic is less important than the processing itself. With client-side and service-side handlers in place, the sample application shows the basics of the JWS handler framework on the sender side and on the intermediary/receiver side.

The UUIDValidator (see Example 3-7) validates an incoming message on the service side. The validator needs access to the entire SOAP message rather than just its body; hence, the validator must be implemented as a SOAPMessageHandler rather than as a LogicalMessageHandler.

Example 3-7. A service-side handler for request validation

```java
package ch03.fib;

import java.util.UUID;
import java.util.Set;
import java.util.Iterator;
import java.util.Locale;
import javax.xml.namespace.QName;
import javax.xml.soap.SOAPMessage;
import javax.xml.soap.SOAPConstants;
import javax.xml.ws.handler.MessageContext;
import javax.xml.ws.handler.soap.SOAPHandler;
import javax.xml.ws.handler.soap.SOAPMessageContext;
import javax.xml.soap.SOAPEnvelope;
import javax.xml.soap.SOAPHeader;
import javax.xml.soap.SOAPBody;
import javax.xml.soap.SOAPHeaderElement;
import javax.xml.soap.SOAPException;
import javax.xml.soap.Node;
import javax.xml.ws.soap.SOAPFaultException;
import javax.xml.soap.SOAPFault;
import java.io.IOException;

public class UUIDValidator implements SOAPHandler<SOAPMessageContext> {
    private static final boolean trace = false; // make true to see message

    public boolean handleMessage(SOAPMessageContext ctx) {
        Boolean response_p = (Boolean)
            ctx.get(MessageContext.MESSAGE_OUTBOUND_PROPERTY);

        // Handle the SOAP only if it's incoming.
        if (!response_p) {
            try {
                SOAPMessage msg = ctx.getMessage();
                SOAPEnvelope env = msg.getSOAPPart().getEnvelope();
                SOAPHeader hdr = env.getHeader();

                // Ensure that the SOAP message has a header.
                if (hdr == null)
                    generateSOAPFault(msg, "No message header.");

                // Get UUID value from header block if it's there.
                Iterator it =
                    hdr.extractHeaderElements(SOAPConstants.URI_SOAP_ACTOR_NEXT);
```

```
            if (it == null || !it.hasNext())
                generateSOAPFault(msg, "No header block for next actor.");
            Node next = (Node) it.next();
            String value = (next == null) ? null : next.getValue();
            if (value == null)
                generateSOAPFault(msg, "No UUID in header block.");

            // Reconstruct a UUID object to check some properties.
            UUID uuid = UUID.fromString(value.trim());
            if (uuid.variant() != UUIDvariant ||
                uuid.version() != UUIDversion)
                generateSOAPFault(msg, "Bad UUID variant or version.");

            if (trace) msg.writeTo(System.out);
        }
        catch(SOAPException e) { System.err.println(e); }
        catch(IOException e) { System.err.println(e); }
    }
    return true; // continue down the chain
}

public boolean handleFault(SOAPMessageContext ctx) {
    return true;
}
public Set<QName> getHeaders() { return null; }
public void close(MessageContext messageContext) { }

private void generateSOAPFault(SOAPMessage msg, String reason) {
    try {
        SOAPBody body = msg.getSOAPPart().getEnvelope().getBody();
        SOAPFault fault = body.addFault();
        fault.setFaultString(reason);
        // wrapper for a SOAP 1.1 or SOAP 1.2 fault
        throw new SOAPFaultException(fault);
    }
    catch(SOAPException e) { }
}
private static final int UUIDvariant = 2; // layout
private static final int UUIDversion = 4; // version
}
```

First, the validator checks whether the incoming message even has a header and, if so, whether the header has the appropriate header block, in this case a header block marked in the SOAP role of next. Here is the code segment, without most of the comments, for the first validation check:

```
if (hdr == null) // hdr refers to the SOAPHeader
    generateSOAPFault(msg, "No message header.");
Iterator it = hdr.extractHeaderElements(SOAPConstants.URI_SOAP_ACTOR_NEXT);
if (it == null || !it.hasNext())
    generateSOAPFault(msg, "No header block for next actor.");
Node next = (Node) it.next();
String value = (next == null) ? null : next.getValue();
if (value == null)
    generateSOAPFault(msg, "No UUID in header block.");
```

If the first validation check succeeds, the validator then checks whether the UUID has the expected properties. In this example, the validator checks for the UUID version number, which should be 4, and the variant number, which should be 2. (The variant controls the format of the UUID value as a string.) Here is the code segment:

```
UUID uuid = UUID.fromString(value.trim());
if (uuid.variant() != UUIDvariant || uuid.version() != UUIDversion)
    generateSOAPFault(msg, "Bad variant or version.");
```

If any step in the validation fails, the validator throws a `SOAPFaultException`, which is a wrapper for SOAP 1.1 and SOAP 1.2 faults. This fault wrapper lets us ignore differences between the two SOAP versions. Here is the code segment that generates the fault:

```
private void generateSOAPFault(SOAPMessage msg, String reason) {
    try {
        SOAPBody body = msg.getSOAPPart().getEnvelope().getBody();
        SOAPFault fault = body.addFault();
        fault.setFaultString(reason);
        throw new SOAPFaultException(fault);
    }
    catch(SOAPException e) { }
}
```

The `UUIDValidator` handler is annotated with `@HandlerChain` and deployed through a configuration file. Here is the code segment that shows the revision to the `RabbitCounter` SIB:

```
@HandlerChain(file = "handler-chain.xml")
public class RabbitCounter {
```

The files *handler-chain.xml* and *UUIDValidator.class* reside in the *ch03/fib* subdirectory of the working directory.

Various tests might be run against the validator. To begin, the line:

```
helem.addTextNode(uuid.toString());
```

in `fibC.UUIDHandler` could be commented out so that the UUID value is not inserted into the header. In this case, the following fault message is generated and picked up by the `handleFault` method in the `UUIDValidator` handler:

```
<S:Envelope xmlns:S="http://schemas.xmlsoap.org/soap/envelope/">
  <S:Header/>
    <S:Body>
      <ns2:Fault xmlns:ns2="http://schemas.xmlsoap.org/soap/envelope/"
                 xmlns:ns3="http://www.w3.org/2003/05/soap-envelope">
        <faultcode>ns2:Server</faultcode>
        <faultstring>No UUID in header block.</faultstring>
        ...
        <message>No UUID in header block.</message>
        ...
      </ns2:Fault>
    </S:Body>
</S:Envelope>
```

The full fault message is large and includes a stack trace.

A more subtle test is to provide a UUID that differs in the expected properties. For this test, the assignment to the `uuid` object reference in `UUIDHandler` on the client side can be changed from:

```
UUID uuid = UUID.randomUUID();
```

to:

```
uuid = new UUID(new java.util.Random().nextLong(),    // lower 64 bits
                new java.util.Random().nextLong());   // upper 64 bits
```

The two-argument constructor generates a UUID with different properties than the `randomUUID` method does, a difference that the validator recognizes. With this change the fault message reads:

```
Bad UUID variant or version.
```

Summary of the Handler Methods

The `Handler` interface declares three methods: `handleMessage`, `handleFault`, and `close`. The `SOAPHandler` extension of this interface adds one more declaration: `getHeaders`, which returns a set of the `QNames` for SOAP header blocks. This method could be used, for example, to insert security information into a SOAP header, as an example in Chapter 5 shows. In the current example, the method simply returns `null`. The other three methods have a `MessageContext` parameter that provides access to the underlying SOAP message.

To illustrate the flow of control, here is the *handler-chain.xml* configuration file for the client-side handlers that execute when the `FibClient` is invoked:

```
<?xml version="1.0" encoding="UTF-8" standalone="yes"?>
<javaee:handler-chains
     xmlns:javaee="http://java.sun.com/xml/ns/javaee"
     xmlns:xsd="http://www.w3.org/2001/XMLSchema">
  <javaee:handler-chain>
    <javaee:handler>
      <javaee:handler-class>fibC.TestHandler</javaee:handler-class>
    </javaee:handler>
    <javaee:handler>
      <javaee:handler-class>fibC.UUIDHandler</javaee:handler-class>
    </javaee:handler>
    <javaee:handler>
      <javaee:handler-class>fibC.ArgHandler</javaee:handler-class>
    </javaee:handler>
  </javaee:handler-chain>
</javaee:handler-chains>
```

The `TestHandler` and `UUIDHandler` are `SOAPHandler`s and the `ArgHandler` is a `LogicalHandler`. Recall that, for outgoing messages such as requests from the `FibClient`, the logical handlers execute before the SOAP handlers. In this example, the order of execution is:

1. The getHeaders method in the TestHandler and then in the UUIDHandler, both of which are SOAP handlers, execute first.

2. The handleMessage in the logical handler ArgHandler executes first among the methods with this name.

3. The handleMessage in the SOAP handler TestHandler executes next, as this handler is listed first in the configuration file.

4. The handleMessage in the SOAP handler UUIDHandler executes next.

5. The close method in the SOAP handler UUIDHandler executes to signal that UUID Handler processing is now done.

6. The close method in the SOAP handler TestHandler executes to signal that the TestHandler processing is now done.

7. The close method in the logical handler ArgHandler executes, which completes the chain on the outgoing message.

The JWS handler framework gives the programmer hooks into the message-processing pipeline so that incoming and outgoing messages can be intercepted and processed in an application-suitable way. The callback handleMessage is especially convenient, providing access to either the entire SOAP message (in a SOAPMessageHandler) or to the payload of a SOAP message (in a LogicalMessageHandler).

The RabbitCounter As a SOAP 1.2 Service

It takes just a few steps to transform the RabbitCounter from a SOAP 1.1 to a SOAP 1.2 service. The critical step is to add a @BindingType annotation to the SIB:

```
import javax.xml.ws.BindingType;
@BindingType(value = "http://java.sun.com/xml/ns/jaxws/2003/05/soap/bindings/HTTP/")
public class RabbitCounter {
```

JWS does have a constant:

```
javax.xml.ws.soap.SOAPBinding.SOAP12HTTP_BINDING
```

for the standard SOAP 1.2 binding, but the Endpoint publisher cannot generate a WSDL using the standard binding. The workaround is to use the nonstandard binding value shown above. When the service is published, Endpoint issues the nonfatal warning:

```
com.sun.xml.internal.ws.server.EndpointFactory generateWSDL
WARNING: Generating non-standard WSDL for the specified binding
```

The *wsgen* utility is used in the same way as before, but the *wsimport* utility is now invoked with the -extension flag to indicate that the client stubs are generated from a nonstandard WSDL:

```
% wsimport -keep -extension -p fibC2 http://localhost:8888/fib?wsdl
```

The UUIDHandler can be changed to exhibit a SOAP 1.2 feature, namely, use of the mustUnderstand attribute in a SOAP 1.2 header. Here is the change, immediately below the setActor call from the SOAP 1.1 version:

```
helem.setActor(SOAPConstants.URI_SOAP_ACTOR_NEXT);
helem.setMustUnderstand(true);  // SOAP 1.2
```

Here is the resulting request message, with the inserted header block in bold:

```
<S:Envelope xmlns:S="http://www.w3.org/2003/05/soap-envelope">
   <S:Header>
     <uuid xmlns="http://ch03.fib"
           xmlns:env="http://www.w3.org/2003/05/soap-envelope"
           env:mustUnderstand="true"
           env:role="http://schemas.xmlsoap.org/soap/actor/next">
        b50064b5-24e8-42bc-9716-10537c86bbd8
     </uuid>
   </S:Header>
   <S:Body>
      <ns2:countRabbits xmlns:ns2="http://ch03.fib">
         <arg0>-45</arg0>
      </ns2:countRabbits>
   </S:Body>
</S:Envelope>
```

The mustUnderstand makes explicit that any intermediary node and the final receiver are expected to process the header block in an application-appropriate way. Even in SOAP 1.1, of course, such processing may occur.

The binding section in the Endpoint-generated WSDL reflects the change from the SOAP 1.1 to SOAP 1.2:

```
<binding name="RabbitCounterPortBinding" type="tns:RabbitCounter">
   <soap12:binding transport="http://www.w3.org/2003/05/soap/bindings/HTTP/"
                    style="document"/>
   <operation name="countRabbits">
     <soap12:operation soapAction=""/>
       <input>
          <soap12:body use="literal"/>
       </input>
       <output>
          <soap12:body use="literal"/>
       </output>
       <fault name="FibException">
          <soap12:fault name="FibException" use="literal"/>
       </fault>
   </operation>
</binding>
```

Given the relatively minor differences between SOAP 1.1 and SOAP 1.2 and the status of SOAP 1.1 as the *de facto* standard, it makes sense to stick with SOAP 1.1 unless there is a compelling reason to use SOAP 1.2. Finally, SOAP 1.2 is backward compatible with SOAP 1.1.

The MessageContext and Transport Headers

This section considers how the JWS level of a service interacts with the transport level. The focus is on the MessageContext, which is normally accessed in handlers: the subtypes SOAPMessageContext and LogicalMessageContext are the parameter types, for example, in the handleMessage callbacks of SOAP and logical handlers, respectively.

The notion of *context* is a familiar one in modern programming systems, including Java. Servlets have a ServletContext, EJBs have an EJBContext (with appropriate subtypes such as SessionContext), and web services have a WebServiceContext. Seen in an architectural light, a context is what gives an object (a servlet, an EJB, a web service) access to its underlying container (servlet container, EJB container, web service container). Containers, in turn, provide the under-the-hood support for the object. Seen in a programming light, a context is a Map<String, Object>, that is, a key/value collection in which the keys are strings and the values are arbitrary objects.

It makes sense that the application level of a @WebService (that is, the SEI and the SIB) usually take the underlying MessageContext for granted, treating it as unseen infrastructure. At the handler level, the MessageContext is appropriately exposed as the data type of callback parameters so that a SOAP or a logical handler can access the SOAP messages and their payloads, respectively. This section examines the more unusual situation in which the MessageContext is accessed outside of handlers; that is, in the application's main components: the service implementation bean (SIB) and its clients.

SOAP messages are delivered predominantly over HTTP. At issue, then, is how much of the HTTP infrastructure is exposed through the MessageContext in Java-based web services. What holds for HTTP also holds for alternative transports such as SMTP or even JMS.

In a handler or SIB, Java provides access to HTTP messages in a MessageContext. In a Java-based client, Java likewise gives access to the HTTP level but in this case through the BindingProvider and the request/response contexts, which are exposed as BindingProvider properties. The code examples illustrate application-level as opposed to handler-level access to transport messages.

An Example to Illustrate Transport-Level Access

The Echo service in Example 3-8 merely echoes a client's text message back to the client. It is best to keep the service simple so that focus is on the transport layer. Throughout the example, the assumption is that transport is SOAP over HTTP.

Example 3-8. A service with access to the message context

```
package ch03.mctx;

import java.util.Map;
import java.util.Set;
import javax.annotation.Resource;
```

```
import javax.jws.WebService;
import javax.jws.WebMethod;
import javax.xml.ws.WebServiceContext;
import javax.xml.ws.handler.MessageContext;
import javax.jws.HandlerChain;

/**
 * A minimalist service to explore the MessageContext.
 * The operation takes a string and echoes it together
 * with transport information back to the client.
 */
@WebService
@HandlerChain(file = "handler-chain.xml")
public class Echo {
    // Enable 'dependency injection' on web service context
    @Resource
    WebServiceContext ws_ctx;

    @WebMethod
    public String echo(String from_client) {
        MessageContext ctx = ws_ctx.getMessageContext();
        Map req_headers = (Map) ctx.get(MessageContext.HTTP_REQUEST_HEADERS);
        MapDump.dump_map((Map) ctx, "");
        String response = "Echoing your message: " + from_client;
        return response;
    }
}
```

The Echo class has a field named ws_ctx of type WebServiceContext, which is annotated
with @Resource. This annotation is used to request *dependency injection*, a notion associated with *AOP* (Aspect-Oriented Programming). As with any uninitialized field,
ws_ctx has a default value of null. Yet, in the first line of the echo method, the method
getMessageContext is invoked on the object to which ws_ctx now, as if by magic, refers.
There is no magic, of course. The JWS container injects a WebServiceContext object
into the application and makes ws_ctx refer to this object. The WebServiceContext then
is used to access the MessageContext, which in turn is used to get a Map of the underlying
transport (typically HTTP) headers. The dump_map method in the utility class MapDump:

```
package ch03.mctx;

import java.util.Map;
import java.util.Set;
public class MapDump {
    public static void dump_map(Map map, String indent) {
        Set keys = map.keySet();
        for (Object key : keys) {
            System.out.println(indent + key + " ==> " + map.get(key));
            if (map.get(key) instanceof Map)
                dump_map((Map) map.get(key), indent += "    ");
        }
    }
}
```

is then invoked to dump the HTTP headers. Example 3-9 shows the relevant part of the dump from a sample client call against the Echo service.

Example 3-9. A dump of a sample HTTP message context

```
javax.xml.ws.wsdl.port ==> {http://mctx.ch03/}EchoPort
javax.xml.ws.soap.http.soapaction.uri ==> "echo"
com.sun.xml.internal.ws.server.OneWayOperation ==> null
javax.xml.ws.http.request.pathinfo ==> null
com.sun.xml.internal.ws.api.message.packet.outbound.transport.headers ==>
  com.sun.net.httpserver.Headers@0
   com.sun.xml.internal.ws.client.handle ==> null
   javax.xml.ws.wsdl.service ==> {http://mctx.ch03/}EchoService
   javax.xml.ws.reference.parameters ==> [ ]
   javax.xml.ws.http.request.headers ==>
      sun.net.httpserver.UnmodifiableHeaders@2c47bd03
      Host ==> [localhost:9797]
      Content-type ==> [text/xml; charset=utf-8]
      Accept-encoding ==> [gzip]
      Content-length ==> [198]
      Connection ==> [keep-alive]
      Greeting ==> [Hello, world!]
      User-agent ==> [Java/1.6.0_06]
      Accept ==> [text/xml, multipart/related, text/html, image/gif,
                 image/jpeg, *; q=.2, */*; q=.2]
      Soapaction ==> ["echo"]
      ...
javax.xml.ws.http.request.method ==> POST
```

Near the end in bold are the HTTP request headers from the client's invocation of the echo operation. On the client side, the underlying libraries insert some standard key/ value pairs into the HTTP headers; for instance, the pairs:

```
Host ==> [localhost:9797]
Content-type ==> [text/xml; charset=utf-8]
```

In standard HTTP, a colon separates the key from the value; in the Java rendering, the arrow ==> is used for clarity. In this example, two of the entries are from an Echo client:

```
Accept-encoding ==> [gzip]
Greeting ==> [Hello, world!]
```

The first pair uses a standard HTTP key, Accept-encoding, with a value of gzip to signal that the client is willing to accept a gzip-compressed SOAP envelope as the body of the HTTP response. The second key/value pair is whimsical: it uses the nonstandard key Greeting. HTTP 1.1, the current version, allows arbitrary key/value pairs to be inserted into the HTTP header.

Another entry in the HTTP header merits special attention:

```
Soapaction ==> ["echo"]
```

For a Java client, the default value for the Soapaction attribute would be the empty string. The string echo occurs here as the value because the client inserts this value into the header of the HTTP request.

The JWS runtime processes the transport-layer headers that client or service code inserts. In this example, the EchoClient shown below uses the key Accept-Encoding (uppercase *E* in Encoding), which the JWS runtime changes to Accept-encoding (lowercase *e* in encoding). Example 3-10 shows the sample client.

Example 3-10. A client that manipulates the message context

```
import java.util.Map;
import java.util.Set;
import java.util.List;
import java.util.Collections;
import java.util.HashMap;
import javax.xml.ws.BindingProvider;
import javax.xml.ws.handler.MessageContext;
import echoC.EchoService;
import echoC.Echo;

class EchoClient {
    public static void main(String[ ] args) {
        EchoService service = new EchoService();
        Echo port = service.getEchoPort();

        Map<String, Object> req_ctx = ((BindingProvider) port).getRequestContext();

        // Sample invocation:
        //
        // % java EchoClient http://localhost:9797 echo
        //
        // 1st command-line argument ends with service location port number
        // 2nd command-line argument is the service operation
        if (args.length >= 2) {
            // Endpoint address becomes: http://localhost:9797/echo
            req_ctx.put(BindingProvider.ENDPOINT_ADDRESS_PROPERTY,
                            args[0] + "/" + args[1]);
            // SOAP action becomes: echo
            req_ctx.put(BindingProvider.SOAPACTION_URI_PROPERTY, args[1]);
        }
        // Add some application-specific HTTP headers
        Map<String, List<String>> my_header = new HashMap<String, List<String>>();
        my_header.put("Accept-Encoding", Collections.singletonList("gzip"));
        my_header.put("Greeting", Collections.singletonList("Hello, world!"));

        // Insert customized headers into HTTP message headers
        req_ctx.put(MessageContext.HTTP_REQUEST_HEADERS, my_header);

        dump_map(req_ctx, "");
        System.out.println("\n\nRequest above, response below\n\n");

        // Invoke service operation to generate an HTTP response.
        String response = port.echo("Have a nice day :)");
```

```
        Map<String, Object> res_ctx = ((BindingProvider) port).getResponseContext();
        dump_map(res_ctx, "");

        Object response_code = res_ctx.get(MessageContext.HTTP_RESPONSE_CODE);
    }

    private static void dump_map(Map map, String indent) {
        Set keys = map.keySet();
        for (Object key : keys) {
            System.out.println(indent + key + " ==> " + map.get(key));
            if (map.get(key) instanceof Map)
                dump_map((Map) map.get(key), indent += "   ");
        }
    }
}
```

Here is the client invocation that resulted in the dump just shown:

```
% java EchoClient http://localhost:9797 echo
```

The two command-line arguments are concatenated in the EchoClient application to generate the service's endpoint location. The second command-line argument, the name of the operation echo, also is inserted into the HTTP request header as the value of SOAPACTION_URI_PROPERTY.

The EchoClient application gains access to the HTTP request headers through the port object reference, which is first cast to a BindingProvider so that the method getRequestContext can be invoked. This method returns a Map into which two entries are made: the first, as just noted, sets the SOAPACTION_URI_PROPERTY to the string echo; the second entry sets the ENDPOINT_ADDRESS_PROPERTY to the URL *http://localhost:9797/ echo*, thus illustrating how the service's endpoint location could be set dynamically in a client application.

The client also creates an empty map, with my_header as the reference, that accepts String keys and Object values. The Object values are of subtype List<String>, instances of which the Collections utility method singletonList generates. The extra key/value pairs are inserted into the HTTP request header. Although the service operation echo is invoked only once in the client, every invocation thereafter would result in an HTTP request with the augmented headers.

Near the end of the EchoClient application, the port object reference is cast again to BindingProvider so that the method getResponseContext can be invoked to gain access to the HTTP response context, which is also a Map. The MessageContext class has various constants such as HTTP_RESPONSE_CODE, which are helpful in extracting information from the HTTP response headers.

The Echo service includes the SOAP handler shown Example 3-11. It is noteworthy that the service, the sample client, and the SOAP handler all have access to the underlying message context, although the syntax for accessing this context differs slightly among the three.

Example 3-11. A service-side handler with access to the message context

```
package ch03.mctx;

import java.util.Map;
import java.util.Set;
import java.util.Locale;
import javax.xml.namespace.QName;
import javax.xml.ws.handler.MessageContext;
import javax.xml.ws.handler.soap.SOAPHandler;
import javax.xml.ws.handler.soap.SOAPMessageContext;

public class EchoHandler implements SOAPHandler<SOAPMessageContext> {
    public boolean handleMessage(SOAPMessageContext ctx) {
        // Is this an inbound message, i.e., a request?
        Boolean response_p = (Boolean)
            ctx.get(MessageContext.MESSAGE_OUTBOUND_PROPERTY);

        // Manipulate the SOAP only if it's incoming.
        if (!response_p) MapDump.dump_map((Map) ctx, "");
        return true; // continue down the chain
    }
    public boolean handleFault(SOAPMessageContext ctx) { return true; }
    public Set<QName> getHeaders() { return null; }
    public void close(MessageContext messageContext) { }
}
```

Later examples return to the theme of application-level access to the transport level, in particular to HTTP or equivalent message headers. For example, HTTP headers are one place to store credentials in a service that requires authentication from its clients.

Web Services and Binary Data

In the examples so far, the underlying SOAP messages contain *text* that is converted, as needed, to service-appropriate types. The type conversion is mostly automatic, occurring in the JWS infrastructure without application intervention. For instance, here is the body of a SOAP request message to the countRabbits operation. The argument 45 occurs as text in the message:

```
<S:Body>
    <ns2:countRabbits xmlns:ns2="http://ch03.fib">
        <arg0>45</arg0>
    </ns2:countRabbits>
</S:Body>
```

but is converted automatically to an int so that the service method countRabbits:

```
@WebMethod
public int countRabbits(int n) throws FibException {
```

can compute and return the Fibonacci number for integer argument n. Neither the FibClient application nor the RabbitCounter service does any explicit type conversion.

By contrast, some explicit type conversions occur at the handler level. For instance, the UUIDHandler on the client side and the UUIDValidator on the service side do explicit, if simple, type conversions. The UUIDHandler converts the UUID object to a string:

```
helem.addTextNode(uuid.toString());
```

and the UUIDValidator does the opposite conversion:

```
UUID uuid = UUID.fromString(value.trim());
```

The client-side ArgHandler does the most work with respect to type conversion. This logical handler uses JAX-B in this code segment:

```
JAXBContext jaxb_ctx = JAXBContext.newInstance("fibC");
Object payload = msg.getPayload(jaxb_ctx);
if (payload instanceof JAXBElement) {
    Object obj = ((JAXBElement) payload).getValue();
    CountRabbits obj_cr = (CountRabbits) obj;
    int n = obj_cr.getArg0();          // current value
    if (n < 0) {                       // negative argument?
        obj_cr.setArg0(Math.abs(n));  // make non-negative

        // Update the message.
        ((JAXBElement) payload).setValue(obj_cr);
        msg.setPayload(payload, jaxb_ctx);
    }
}
```

to obtain a CountRabbits object from an XML element; invokes the getArg0 method and possibly the setArg0 method on the object to ensure that the argument passed to the countRabbits operation is nonnegative; and then, with the call to setPayload, changes the CountRabbits object back to an XML element.

Type conversions come to the forefront in the issue of how binary data such as images, movies, and the like can be arguments passed to or values returned from web service operations. SOAP-based web services are not limited to text, but their use of binary data raises important efficiency issues.

There are two general approaches to handling arbitrary binary data in SOAP-based web services:

- The binary data can be encoded using a scheme such as base64 and then transmitted as the payload of the SOAP body. For instance, a service operation that returns an image to a requester simply could return a java.awt.Image, which is a Java wrapper for image bytes. The image's bytes then would be encoded and transmitted as the body of a SOAP message. The downside is that base64 and similar encoding schemes result in payloads that are at least a third larger in size than the original, unencoded binary data. In short, byte encoding such as base64 results in data bloat.

- The binary data can be transmitted as one or more attachments to a SOAP message. Recall that a SOAP message consists of a SOAP part, which is the SOAP envelope with an optional header and a possibly empty body. A SOAP message also may have attachments, which can carry data of any MIME type, including multimedia types such as `audio/x-wav`, `video/mpeg`, and `image/jpeg`. JAX-B provides the required mappings between MIME and Java types: the MIME types `image/*` map to `Image`, and the remaining multimedia types map to `DataHandler`.

The attachments option is preferable because it avoids data bloat—raw rather than encoded bytes go from sender to receiver. The downside is that the receiver then must deal with the raw bytes, for example, by converting them back into multimedia types such as images and sounds.

Three Options for SOAP Attachments

There are basically three options for SOAP attachments: *SwA* (SOAP with Attachments), the original SOAP specification for attachments; *DIME* (Direct Internet Message Encapsulation), a lightweight but by now old-fashioned encoding scheme for attachments; and *MTOM* (Message Transmission Optimization Mechanism), which is based on *XOP* (XML-Binary Optimized Packaging). JWS has a DIME extension whose main purpose is to interoperate with Microsoft clients. Up until the release of Microsoft Office 2003, a web service client written in Visual Basic for Applications (VBA) could handle only DIME rather than MTOM attachments. The SwA approach has drawbacks. For one, it is hard to use SwA with a `document`-style service, which is now the norm. Further, frameworks such as DotNet do not support SwA. MTOM has the W3C stamp of approval and enjoys widespread support; hence, MTOM is the efficient, modern, interoperable way to transmit binary data in SOAP-based web services. Before considering MTOM, let's take a quick look at base64 encoding of binary data, which might be used for small binary payloads.

Using Base64 Encoding for Binary Data

The `SkiImageService` in Example 3-12 has two operations: `getImage` returns a named image of a skier and `getImages` returns a list of the available images. Example 3-12 shows the source code.

Example 3-12. A service that provides images as responses

```
package ch03.image;

import javax.jws.WebService;
import javax.jws.WebMethod;
import java.util.Map;
import java.util.HashMap;
import java.util.Set;
import java.util.List;
import java.util.ArrayList;
import java.util.Iterator;
import java.awt.Image;
```

```java
import java.io.FileInputStream;
import java.io.ByteArrayOutputStream;
import java.io.ByteArrayInputStream;
import java.io.IOException;
import javax.imageio.ImageIO;
import javax.imageio.stream.ImageInputStream;
import javax.imageio.ImageReader;
import javax.jws.HandlerChain;

@WebService(serviceName = "SkiImageService")
@HandlerChain(file = "handler-chain.xml") // for message tracking
public class SkiImageService {
    // Returns one image given the image's name.
    @WebMethod
    public Image getImage(String name) { return createImage(name);   }

    // Returns a list of all available images.
    @WebMethod
    public List<Image> getImages() { return createImageList(); }

    public SkiImageService() {
        photos = new HashMap<String, String>();
        photos.put("nordic", "nordic.jpg");
        photos.put("alpine", "alpine.jpg");
        photos.put("telemk", "telemk.jpg");
        default_key = "nordic";
    }

    // Create a named image from the raw bytes.
    private Image createImage(String name) {
        byte[ ] bytes = getRawBytes(name);
        ByteArrayInputStream in = new ByteArrayInputStream(bytes);
        Iterator iterators = ImageIO.getImageReadersByFormatName("jpeg");
        ImageReader iterator = (ImageReader) iterators.next();
        try {
            ImageInputStream iis = ImageIO.createImageInputStream(in);
            iterator.setInput(iis, true);
            return iterator.read(0);
        }
        catch(IOException e) {
            System.err.println(e);
            return null;
        }
    }

    // Create a list of all available images.
    private List<Image> createImageList() {
        List<Image> list = new ArrayList<Image>();
        Set<String> key_set = photos.keySet();
        for (String key : key_set) {
            Image image = createImage(key);
            if (image != null) list.add(image);
        }
        return list;
    }
```

```
// Read the bytes from the file for one image.
private byte[ ] getRawBytes(String name) {
    ByteArrayOutputStream out = new ByteArrayOutputStream();
    try {
        String cwd = System.getProperty ("user.dir");
        String sep = System.getProperty ("file.separator");
        String base_name = cwd + sep + "jpegs" + sep;
        String file_name = base_name + name + ".jpg";
        FileInputStream in = new FileInputStream(file_name);

        // Send default image if there's none with this name.
        if (in == null) in = new FileInputStream(base_name + "nordic.jpg");
        byte[ ] buffer = new byte[2048];
        int n = 0;
        while ((n = in.read(buffer)) != -1)
            out.write(buffer, 0, n); // append to ByteArrayOutputStream
        in.close();
    }
    catch(IOException e) { System.err.println(e); }
    return out.toByteArray();
}
private static final String[ ] names = {
    "nordic.jpg", "tele.jpg", "alpine.jpg" };
private Map<String, String> photos;
private String default_key;
}
```

Most of the service code is in utility methods that read the image's bytes from a local
file and then create an Image from these bytes. The images are stored in a subdirectory
of the working directory. The two service operations, getImage and getImages, are un-
complicated. The getImage operation returns a java.awt.Image, and the getImages operation returns a List<Image>. Of interest here is that the return types are high level
rather than byte arrays.

A quick look at the binding section of the WSDL:

```
<binding name="SkiImageServicePortBinding" type="tns:SkiImageService">
    <soap:binding transport="http://schemas.xmlsoap.org/soap/http"
                style="document"></soap:binding>
    <operation name="getImage">
        <soap:operation soapAction=""></soap:operation>
            <input>
                <soap:body use="literal"></soap:body>
            </input>
            <output>
                <soap:body use="literal"></soap:body>
            </output>
    </operation>
    <operation name="getImages">
        <soap:operation soapAction=""></soap:operation>
            <input>
                <soap:body use="literal"></soap:body>
            </input>
```

```
      <output>
        <soap:body use="literal"></soap:body>
      </output>
    </operation>
  </binding>
```

shows that the service is document style with literal encoding. Two segments from the associated XSD provide more information:

```
<xs:complexType name="getImagesResponse">
  <xs:sequence>
    <xs:element name="return" type="xs:base64Binary"
                minOccurs="0" maxOccurs="unbounded"></xs:element>
  </xs:sequence>
</xs:complexType>
```

```
<xs:complexType name="getImageResponse">
  <xs:sequence>
    <xs:element name="return" type="xs:base64Binary" minOccurs="0"></xs:element>
  </xs:sequence>
</xs:complexType>
```

The XSD indicates that the style is indeed wrapped document, with getImageResponse as one of the wrapper types. The XSD type for this wrapper is the expected base64Binary.

Here is the truncated SOAP response envelope from a request for one of the skiing images:

```
<S:Envelope xmlns:S="http://schemas.xmlsoap.org/soap/envelope/">
  <S:Body>
    <ns2:getImageResponse xmlns:ns2="http://image.ch03/">
      <return>iVBORw0KGgoAAAANSUhEUgAAAZAAAAEsCAIAAABi1X...</return>
    </ns2:getImageResponse>
  </S:Body>
</S:Envelope>
```

The entire image or list of images is returned as a base64 character encoding in the body of the SOAP envelope.

The SkiImageClient is shown here:

```
import skiC.SkiImageService_Service;
import skiC.SkiImageService;
import java.util.List;
class SkiImageClient {
    public static void main(String[ ] args) {
        // wsimport-generated artifacts
        SkiImageService_Service service = new SkiImageService_Service();
        SkiImageService port = service.getSkiImageServicePort();
        // Note the return types: byte[ ] and List<byte[ ]>
        byte[ ] image = port.getImage("nordic");
        List<byte[ ]> images = port.getImages();
        /* Transform the received bytes in some useful way :) */
    }
}
```

This client uses *wsimport*-generated stubs to invoke the service operations. However, the client gets either an array of bytes or a list of these as return values because the XSD type base64Binary maps to the Java type byte[]. The client receives the base64 encoding of images as byte arrays and then must transform these encodings back into images. This is inconvenient, to say the least.

The fix is to edit the WSDL so that the service is friendlier to clients. After saving the WSDL and its XSD to local files (for instance, *ski.wsdl* and *ski.xsd*), the following changes should be made:

1. Edit the XSD document. In particular, make the two additions shown in bold:

```
<xs:complexType name="getImagesResponse">
  <xs:sequence>
    <xs:element name="return" type="xs:base64Binary"
                minOccurs="0" maxOccurs="unbounded"
                xmime:expectedContentTypes="image/jpeg"
                  xmlns:xmime="http://www.w3.org/2005/05/xmlmime">
    </xs:element>
  </xs:sequence>
</xs:complexType>

<xs:complexType name="getImageResponse">
  <xs:sequence>
    <xs:element name="return" type="xs:base64Binary"
                minOccurs="0"
                xmime:expectedContentTypes="image/jpeg"
                  xmlns:xmime="http://www.w3.org/2005/05/xmlmime">
    </xs:element>
  </xs:sequence>
</xs:complexType>
```

The attribute expectedContentTypes is set to the MIME type image/jpeg so that the *wsimport* utility can generate versions of the operations getImage and getImages that return the Java types Image and List<Image>, respectively.

2. In the SkiImageService, the @WebService annotation should be changed to:

```
@WebService(serviceName = "SkiImageService",
            wsdlLocation = "ch03/image/ski.wsdl")
```

to reflect the new location of the WSDL.

3. The WSDL should then be changed to reflect the new location of XSD document:

```
<types>
  <xsd:schema>
    <xsd:import namespace="http://image.ch03/"
                schemaLocation="ski.xsd">
    </xsd:import>
  </xsd:schema>
</types>
```

4. From the working directory (in this case, the parent of the subdirectory *ch03*), the *wsimport* utility is run to generate the new artifacts:

```
% wsimport -keep -p skiC2 ch03/image/ski.wsdl
```

With these changes, the client now can work with the Java types `Image` and `List<Image>` instead of the bytes from the base64 encoding of images. Here is the revised client:

```java
import skiC2.SkiImageService_Service;
import skiC2.SkiImageService;
import java.awt.Image;
import java.util.List;

class SkiImageClient2 {
    public static void main(String[ ] args) {
        SkiImageService_Service service = new SkiImageService_Service();
        SkiImageService port = service.getSkiImageServicePort();

        Image image = port.getImage("telemk");
        List<Image> images = port.getImages();
        /* Process the images in some appropriate way. */
    }
}
```

The revised `SkiImageService` and the corresponding *wsimport*-generated artifacts let clients receive high-level types such as the `java.awt.Image` instead of arrays of raw bytes. These revisions do not remedy the data bloat of base64 encoding, however. The next subsection gives an example that avoids base64 encoding altogether by working with raw rather than encoded bytes.

Using MTOM for Binary Data

This section adapts the `SkiImageService` to use MTOM. Several things need to be changed, but each change is relatively minor.

In the XSD, the `expectedContentTypes` attribute occurs twice. The changes are in bold:

```xml
<xs:complexType name="getImagesResponse">
  <xs:sequence>
    <xs:element name="return" type="xs:base64Binary"
                minOccurs="0" maxOccurs="unbounded"
                xmime:expectedContentTypes="application/octet-stream"
                  xmlns:xmime="http://www.w3.org/2005/05/xmlmime">
    </xs:element>
  </xs:sequence>
</xs:complexType>

<xs:complexType name="getImageResponse">
  <xs:sequence>
    <xs:element name="return" type="xs:base64Binary"
                minOccurs="0"
                xmime:expectedContentTypes="application/octet-stream"
                  xmlns:xmime="http://www.w3.org/2005/05/xmlmime">
    </xs:element>
  </xs:sequence>
</xs:complexType>
```

The MIME subtype name `application/octet-stream` captures the optimization that recommends MTOM—the image bytes will be streamed unencoded to the service client. The bloat of base64 or comparable encoding is thereby avoided.

For emphasis, the `SkiImageService` is annotated to show that MTOM may come into play:

```
import javax.xml.ws.soap.SOAPBinding;
@WebService(serviceName = "SkiImageService")
// This binding value is enabled by default but put here for emphasis.
@BindingType(value = SOAPBinding.SOAP11HTTP_MTOM_BINDING)
//** @HandlerChain(file = "handler-chain.xml") // disable for MTOM
public class SkiImageService {
```

The binding value is the default, hence optional, but it is included for emphasis.

The next change is important. Here is the revised publisher, with the work now divided among several methods for clarity:

```
package ch03.image;

import javax.xml.ws.Endpoint;
import javax.xml.ws.soap.SOAPBinding;
public class SkiImagePublisher {
    private Endpoint endpoint;

    public static void main(String[ ] args) {
        SkiImagePublisher me = new SkiImagePublisher();
        me.create_endpoint();
        me.configure_endpoint();
        me.publish();
    }

    private void create_endpoint() {
        endpoint = Endpoint.create(new SkiImageService());
    }

    private void configure_endpoint() {
        SOAPBinding binding = (SOAPBinding) endpoint.getBinding();
        binding.setMTOMEnabled(true);
    }

    private void publish() {
        int port = 9999;
        String url = "http://localhost:" + port + "/ski";
        endpoint.publish(url);
        System.out.println(url);
    }
}
```

The service endpoint enables MTOM for responses to the client. Although the service still uses a SOAP handler to dump the response message, this dump is misleading because it shows the images as base64-encoded values in the SOAP body. The MTOM optimization occurs *after* the handler executes; hence, a utility such as *tcpdump* offers a better picture of what is really going on.

The last change is to the sample client, now `SkiImageClient3`, after the *wsimport* utility has been run yet again to generate the artifacts in directory *skiC3*:

```java
import skiC3.SkiImageService_Service;
import skiC3.SkiImageService;
import javax.xml.ws.BindingProvider;
import javax.xml.ws.soap.SOAPBinding;
import java.util.List;
import java.io.IOException;
import javax.activation.DataHandler;

class SkiImageClient3 {
    public static void main(String[ ] args) {
        SkiImageService_Service service = new SkiImageService_Service();
        SkiImageService port = service.getSkiImageServicePort();

        DataHandler image = port.getImage("nordic");
        List<DataHandler> images = port.getImages();
        dump(image);
        for (DataHandler hd : images)
            dump(hd);
    }
    private static void dump(DataHandler dh) {
        System.out.println();
        try {
            System.out.println("MIME type: " + dh.getContentType());
            System.out.println("Content:   " + dh.getContent());
        }
        catch(IOException e) { System.err.println(e); }
    }
}
```

The service operations now return a `DataHandler` or a `List` of these. This change is reflected in the JAX-B artifacts created with the *wsimport* utility. For instance, here is a segment from the `GetImageResponse` artifact with an `@XmlMimeType` annotation that reflects the service's revised XSD and also indicates that the service now returns a `DataHandler`:

```java
@XmlElement(name = "return")
@XmlMimeType("application/octet-stream")
protected DataHandler _return;
```

It is now up to the client to reconstruct the appropriate objects from the optimized byte stream sent from the service. The tradeoff is clear: MTOM optimizes the transmission by avoiding data bloat, but the message receiver is forced to deal with raw bytes. Here is the output from the dump method on a sample run:

```
MIME type: application/octet-stream
Content:   java.io.ByteArrayInputStream@210b5b
```

```
MIME type: application/octet-stream
Content:   java.io.ByteArrayInputStream@170888e

MIME type: application/octet-stream
Content:   java.io.ByteArrayInputStream@11563ff
...
```

The input streams provide access to the underlying bytes from which the Image instances can be restored. The service's createImage method could be used on the client side for this purpose.

If the client needed to send large amounts of binary data to the web service, then the client, too, could enable MTOM. This revised SkiImageClient2 shows how by using the Binding method setMTOMEnable:

```
SkiImageService port = service.getSkiImageServicePort();

// Enable MTOM for client transmissions
BindingProvider bp = (BindingProvider) port;
SOAPBinding binding = (SOAPBinding) bp.getBinding();
binding.setMTOMEnabled(true);
```

MTOM does come at a cost, which the receiver pays by having to deal with the raw bytes available through a returned DataHandler's input stream. Yet the gains in efficiency for large data sets may easily offset this cost. As the example shows, enabling MTOM is relatively straightforward on the service side and on the client side.

What's Next?

There is much to like about SOAP-based web services in general and @WebServices in particular. Such services are built upon industry-standard, vendor-independent protocols such as HTTP, XML, and SOAP itself. These services represent a language-neutral approach to building and deploying distributed software systems. The WSDL and associated tools such as Java's *wsimport* ease the task of writing service clients and even services themselves in some preferred language. SOAP as a language-agnostic messaging system and XML Schema as a language-neutral type system promote service interoperability and API standardization.

Yet SOAP-based services are complicated, especially if the developer has to drill down into the infrastructure for any reason whatsoever. Even at the application level, the APIs for SOAP-based web services have become quite rich. There are various standards bodies involved in SOAP and SOAP-based web services, including the W3C, *OASIS* (Organization for the Advancement of Structured Information Services), *IETF* (International Engineering Task Force), and *WS-I* (Web Services Interoperability Organization). There are specification initiatives in areas as broad and varied as interoperability, business process, presentation, security, metadata, reliability, resources, messaging, XML, management, transactions, and SOAP itself. Each area has subareas. For instance, the security area has 10 subareas, the interoperability area likewise has 10 subareas, the metadata area has 9 subareas, and the messaging area also has 9 subareas.

It is not uncommon to hear complaints about how SOAP and SOAP-based web services have been over-engineered. The JAX-WS framework reflects this complexity with its wealth of annotations and tools. This complexity explains, at least in part, the current popularity of REST-style or RESTful approaches to web services. The next chapter focuses on RESTful Web services.

## CHAPTER 4

RESTful Web Services

What Is REST?

Roy Fielding (*http://roy.gbiv.com*) coined the acronym REST in his Ph.D. dissertation. Chapter 5 of his dissertation lays out the guiding principles for what have come to be known as REST-style or RESTful web services. Fielding has an impressive resume. He is, among other things, a principal author of the HTTP specification and a cofounder of the Apache Software Foundation.

REST and SOAP are quite different. SOAP is a messaging protocol, whereas REST is a style of software architecture for distributed hypermedia systems; that is, systems in which text, graphics, audio, and other media are stored across a network and interconnected through hyperlinks. The World Wide Web is the obvious example of such a system. As our focus is *web* services, the World Wide Web is the distributed hypermedia system of interest. In the Web, HTTP is both a transport protocol and a messaging system because HTTP requests and responses are messages. The payloads of HTTP messages can be typed using the MIME type system, and HTTP provides response status codes to inform the requester about whether a request succeeded and, if not, why.

REST stands for REpresentation State Transfer, which requires clarification because the central abstraction in REST—the *resource*—does not occur in the acronym. A resource in the RESTful sense is anything that has an URI; that is, an identifier that satisfies formatting requirements. The formatting requirements are what make URIs *uniform*. Recall, too, that URI stands for Uniform *Resource* Identifier; hence, the notions of *URI* and *resource* are intertwined.

In practice, a resource is an informational item that has hyperlinks to it. Hyperlinks use URIs to do the linking. Examples of resources are plentiful but likewise misleading in suggesting that resources must have something in common other than identifiability through URIs. The gross national product of Lithuania in 2001 is a resource, as is the Modern Jazz Quartet. Ernie Bank's baseball accomplishments count as a resource, as does the maximum flow algorithm. The concept of a resource is remarkably broad but, at the same time, impressively simple and precise.

As Web-based informational items, resources are pointless unless they have at least one representation. In the Web, representations are MIME-typed. The most common type of resource representation is probably still `text/html`, but nowadays resources tend to have multiple representations. For example, there are various interlinked HTML pages that represent the Modern Jazz Quartet, but there are also audio and audiovisual representations of this resource.

Resources have state. For example, Ernie Bank's baseball accomplishments changed during his career with the Chicago Cubs from 1953 through 1971 and culminated in his 1977 induction into the Baseball Hall of Fame. A useful representation must capture a resource's state. For example, the current HTML pages on Ernie at the Baseball Reference website (*http://www.baseball-reference.com*) need to represent all of his major league accomplishments, from his rookie year in 1953 through his induction into the Hall of Fame.

In a RESTful request targeted at a resource, the resource itself remains on the service machine. The requester typically receives a *representation* of the resource if the request succeeds. It is the representation that transfers from the service machine to the requester machine. In different terms, a RESTful client issues a request that involves a resource, for instance, a request to *read* the resource. If this read request succeeds, a typed representation (for instance, `text/html`) of the resource is transferred from the server that hosts the resource to the client that issued the request. The representation is a good one only if it captures the resource's state in some appropriate way.

In summary, RESTful web services require not just resources to represent but also client-invoked operations on such resources. At the core of the RESTful approach is the insight that HTTP, despite the occurrence of *Transport* in its name, is an API and not simply a transport protocol. HTTP has its well-known *verbs*, officially known as *methods*. Table 4-1 shows the HTTP verbs that correspond to the *CRUD* (Create, Read, Update, Delete) operations so familiar throughout computing.

Table 4-1. HTTP verbs and CRUD operations

HTTP verb	Meaning in CRUD terms
POST	*Create* a new resource from the request data
GET	*Read* a resource
PUT	*Update* a resource from the request data
DELETE	*Delete* a resource

Although HTTP is not case-sensitive, the HTTP verbs are traditionally written in uppercase. There are additional verbs. For example, the verb HEAD is a variation on GET that requests only the HTTP headers that would be sent to fulfill a GET request. There are also TRACE and INFO verbs.

Figure 4-1 is a whimsical depiction of a resource with its identifying URI, together with a RESTful client and some typed representations sent as responses to HTTP requests for the resource. Each HTTP request includes a verb to indicate which CRUD operation should be performed on the resource. A good representation is precisely one that matches the requested operation and captures the resource's state in some appropriate way. For example, in this depiction a GET request could return my biography as a hacker as either an HTML document or a short video summary. The video would fail to capture the state of the resource if it depicted, say, only the major disasters in my brother's career rather than those in my own. A typical HTML representation of the resource would include hyperlinks to other resources, which in turn could be the target of HTTP requests with the appropriate CRUD verbs.

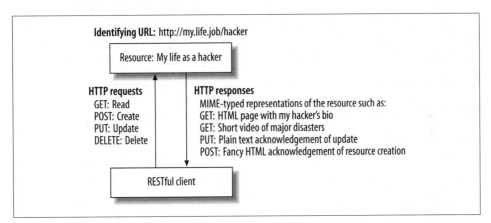

Figure 4-1. A small slice of a RESTful system

HTTP also has standard response codes, such as 404 to signal that the requested resource could not be found, and 200 to signal that the request was handled successfully. In short, HTTP provides request verbs and MIME types for client requests and status codes (and MIME types) for service responses.

Modern browsers generate only GET and POST requests. Moreover, many applications treat these two types of requests interchangeably. For example, Java `HttpServlets` have callback methods such as `doGet` and `doPost` that handle GET and POST requests, respectively. Each callback has the same parameter types, `HttpServletRequest` (the key/value pairs from the requester) and `HttpServletResponse` (a typed response to the requester). It is common to have the two callbacks execute the same code (for instance, by having one invoke the other), thereby conflating the original HTTP distinction between *read* and *create*. A key guiding principle of the RESTful style is to respect the original meanings of the HTTP verbs. In particular, any GET request should be side effect-free (or, in jargon, *idempotent*) because a GET is a *read* rather than a *create*, *update*, or *delete* operation. A GET as a *read* with no side effects is called a *safe GET*.

The REST approach does not imply that either resources or the processing needed to generate adequate representations of them are simple. A REST-style web service might be every bit as subtle and complicated as a SOAP-based service. The RESTful approach tries to simplify matters by taking what HTTP, with its MIME type system, already offers: built-in CRUD operations, uniformly identifiable resources, and typed representations that can capture a resource's state. REST as a design philosophy tries to isolate application complexity at the endpoints, that is, at the client and at the service. A service may require lots of logic and computation to maintain resources and to generate adequate representation of resources—for instance, large and subtly formatted XML documents—and a client may require significant XML processing to extract the desired information from the XML representations transferred from the service to the client. Yet the RESTful approach keeps the complexity out of the transport level, as a resource representation is transferred to the client as the body of an HTTP response message. By contrast, a SOAP-based service inevitably complicates the transport level because a SOAP *message* is encapsulated as the body of a transport message; for instance, an HTTP or SMTP message. SOAP requires messages within messages, whereas REST does not.[*]

Verbs and Opaque Nouns

A URI is meant to be *opaque*, which means that the URI:

```
http://bedrock/citizens/fred
```

has no inherent connection to the URI:

```
http://bedrock/citizens
```

although Fred happens to be a citizen of Bedrock. These are simply two different, independent identifiers. Of course, a good URI designer will come up with URIs that are suggestive about what they are meant to identify. The point is that URIs have no intrinsic hierarchical structure. URIs can and should be interpreted, but these interpretations are imposed on URIs, not inherent in them. Although URI syntax looks like the syntax used to navigate a hierarchical file system, this resemblance is misleading. A URI is an opaque identifier, a logically proper name that denotes exactly one resource.

In RESTful services, then, URIs act as identifying nouns and HTTP methods act as verbs that specify operations on the resources identified by these nouns. For reference, here is the HTTP start line from a client's request against the `TimeServer` service in Chapter 1:

```
POST http://127.0.0.1:9876/ts HTTP/ 1.1
```

The HTTP verb comes first, then the URI, and finally the requester's version of HTTP. This URI is, of course, a URL that locates the web service. Table 4-2 uses simplified URIs to summarize the intended meanings of HTTP/URI combinations.

---

[*] For a thorough coverage of REST-style web services, see Leonard Richardson and Sam Ruby's book *RESTful Web Services* (O'Reilly).

Table 4-2. Sample HTTP verb/URI pairs

HTTP verb/URI	Intended CRUD meaning
POST emps	Create a new employee from the request data
GET emps	Read a list of all employees
GET emps?id=27	Read a singleton list of employee 27
PUT emps	Update the employee list with the request data
DELETE emps	Delete the employee list
DELETE emps?id=27	Delete employee 27

These verb/URI pairs are terse, precise, and uniform in style. The pairs illustrate that RESTful conventions can yield simple, clear expressions about which operation should be performed on which resource. The POST and PUT verbs are used in requests that have an HTTP body; hence, the request data are housed in the HTTP message body. The GET and DELETE verbs are used in requests that have no body; hence, the request data are sent as query string entries.

For the record, RESTful web services are Turing complete; that is, these services are equal in power to any computational system, including a system that consists of SOAP-based web services. Yet the decision about whether to be RESTful in a particular application depends, as always, on practical matters. This first section has looked at REST from on high; it is now time to descend into details through examples.

From @WebService to @WebServiceProvider

The @WebService annotation signals that the messages exchanged between the service and its clients will be SOAP envelopes. The @WebServiceProvider signals that the exchanged messages will be XML documents of some type, a notion captured in the phrase *raw XML*. Of course, a @WebServiceProvider could process and generate SOAP on its own, but this approach is not recommended. (A later example illustrates, however.) The obvious way to provide a SOAP-based web service is to use the annotation @WebService.

In a RESTful request/response service, the service response is raw XML but the incoming request might not be XML at all. A GET request does not have a body; hence, arguments sent as part of the request occur as attributes in the *query string*, a collection of key/value pairs. Here is a sample:

```
http://www.onlineparlor.com/bets?horse=bigbrown&jockey=kent&amount=25
```

The question mark (?) begins the query string, and the attributes are key/value pairs separated by ampersands (&). The order of attributes in the query string is arbitrary; for instance, the `jockey` attribute could occur first in the query string without changing the meaning of the request. By contrast, a POST request does have a body, which can be an arbitrary XML document instead of a SOAP envelope.

A service annotated with @WebServiceProvider implements the Provider interface, which requires that the invoke method:

```
public Source invoke(Source request)
```

be defined. This method expects a Source of bytes (for instance, the bytes in an XML document that represents the service request) and returns a Source of bytes (the bytes in the XML response). When a request arrives, the infrastructure dispatches the request to the invoke method, which handles the request in some service-appropriate way. These points can be illustrated with an example.

A RESTful Version of the Teams Service

The first RESTful service revises the Teams SOAP-based service from Chapter 1. The teams in question are comedy groups such as the Marx Brothers. To begin, the RESTful service honors only GET requests, but the service will be expanded to support the other HTTP verbs associated with the standard CRUD operations.

The WebServiceProvider Annotation

Example 4-1 is the source code for the initial version of the RestfulTeams service.

Example 4-1. The RestfulTeams web service

```
package ch04.team;

import javax.xml.ws.Provider;
import javax.xml.transform.Source;
import javax.xml.transform.stream.StreamSource;
import javax.annotation.Resource;
import javax.xml.ws.BindingType;
import javax.xml.ws.WebServiceContext;
import javax.xml.ws.handler.MessageContext;
import javax.xml.ws.http.HTTPException;
import javax.xml.ws.WebServiceProvider;
import javax.xml.ws.ServiceMode;
import javax.xml.ws.http.HTTPBinding;
import java.io.ByteArrayInputStream;
import java.io.ByteArrayOutputStream;
import java.util.Collections;
import java.util.Map;
import java.util.HashMap;
import java.util.List;
import java.util.ArrayList;
import java.io.IOException;
import java.io.File;
import java.io.FileInputStream;
import java.io.FileOutputStream;
import java.beans.XMLEncoder;
import java.beans.XMLDecoder;
```

```
// The class below is a WebServiceProvider rather than the more usual
// SOAP-based WebService. The service implements the generic Provider
// interface rather than a customized SEI with designated @WebMethods.
@WebServiceProvider

// There are two ServiceModes: PAYLOAD, the default, signals that the service
// wants access only to the underlying message payload (e.g., the
// body of an HTTP POST request); MESSAGE signals that the service wants
// access to entire message (e.g., the HTTP headers and body).
@ServiceMode(value = javax.xml.ws.Service.Mode.MESSAGE)

// The HTTP_BINDING as opposed, for instance, to a SOAP binding.
@BindingType(value = HTTPBinding.HTTP_BINDING)
public class RestfulTeams implements Provider<Source> {
    @Resource
    protected WebServiceContext ws_ctx;

    private Map<String, Team> team_map; // for easy lookups
    private List<Team> teams;           // serialized/deserialized
    private byte[ ] team_bytes;         // from the persistence file

    private static final String file_name = "teams.ser";

    public RestfulTeams() {
        read_teams_from_file(); // read the raw bytes from teams.ser
        deserialize();          // deserialize to a List<Team>
    }

    // This method handles incoming requests and generates the response.
    public Source invoke(Source request) {
        if (ws_ctx == null) throw new RuntimeException("DI failed on ws_ctx.");

        // Grab the message context and extract the request verb.
        MessageContext msg_ctx = ws_ctx.getMessageContext();
        String http_verb = (String)
            msg_ctx.get(MessageContext.HTTP_REQUEST_METHOD);
        http_verb = http_verb.trim().toUpperCase();

        // Act on the verb. To begin, only GET requests accepted.
        if (http_verb.equals("GET")) return doGet(msg_ctx);
        else throw new HTTPException(405); // method not allowed
    }

    private Source doGet(MessageContext msg_ctx) {
        // Parse the query string.
        String query_string = (String) msg_ctx.get(MessageContext.QUERY_STRING);

        // Get all teams.
        if (query_string == null)
            return new StreamSource(new ByteArrayInputStream(team_bytes));
        // Get a named team.
        else {
            String name = get_value_from_qs("name", query_string);
```

```java
            // Check if named team exists.
            Team team = team_map.get(name);
            if (team == null) throw new HTTPException(404); // not found
            // Otherwise, generate XML and return.
            ByteArrayInputStream stream = encode_to_stream(team);
            return new StreamSource(stream);
        }
    }

    private ByteArrayInputStream encode_to_stream(Object obj) {
        // Serialize object to XML and return
        ByteArrayOutputStream stream = new ByteArrayOutputStream();
        XMLEncoder enc = new XMLEncoder(stream);
        enc.writeObject(obj);
        enc.close();
        return new ByteArrayInputStream(stream.toByteArray());
    }

    private String get_value_from_qs(String key, String qs) {
        String[ ] parts = qs.split("=");
        // Check if query string has form: name=<team name>
        if (!parts[0].equalsIgnoreCase(key))
            throw new HTTPException(400); // bad request
        return parts[1].trim();
    }

    private void read_teams_from_file() {
        try {
            String cwd = System.getProperty ("user.dir");
            String sep = System.getProperty ("file.separator");
            String path = get_file_path();
            int len = (int) new File(path).length();
            team_bytes = new byte[len];
            new FileInputStream(path).read(team_bytes);
        }
        catch(IOException e) { System.err.println(e); }
    }

    private void deserialize() {
        // Deserialize the bytes into a list of teams
        XMLDecoder dec = new XMLDecoder(new ByteArrayInputStream(team_bytes));
        teams = (List<Team>) dec.readObject();

        // Create a map for quick lookups of teams.
        team_map = Collections.synchronizedMap(new HashMap<String, Team>());
        for (Team team : teams) team_map.put(team.getName(), team);
    }

    private String get_file_path() {
        String cwd = System.getProperty ("user.dir");
        String sep = System.getProperty ("file.separator");
        return cwd + sep + "ch04" + sep + "team" + sep + file_name;
    }
}
```

The JWS annotations indicate the shift from a SOAP-based to a REST-style service. The main annotation is now @WebServiceProvider instead of @WebService. In the next two annotations:

```
@ServiceMode(value = javax.xml.ws.Service.Mode.MESSAGE)
@BindingType(value = HTTPBinding.HTTP_BINDING)
```

the @ServiceMode annotation overrides the default value of PAYLOAD in favor of the value MESSAGE. This annotation is included only to highlight it, as the RestfulTeams service would work just as well with the default value. The second annotation announces that the service deals with raw XML over HTTP instead of SOAP over HTTP.

The RESTful revision deals with raw XML rather than with SOAP. The comedy teams are now stored on the local disk, in a file named *teams.ser*, as an XML document generated using the XMLEncoder class. Here is a segment of the file:

```
<?xml version="1.0" encoding="UTF-8"?>
<java version="1.6.0_06" class="java.beans.XMLDecoder">
 <object class="java.util.ArrayList">
  <void method="add">
   <object class="ch04.team.Team">
    <void property="name">
     <string>BurnsAndAllen</string>
    </void>
    <void property="players">
     <object class="java.util.ArrayList">
      <void method="add">
       <object class="ch04.team.Player">
        <void property="name">
         <string>George Burns</string>
        </void>
        <void property="nickname">
         <string>George</string>
        </void>
       </object>
      </void>
      ...
</java>
```

An XMLDecoder is used to deserialize this stored XML document into a List<Team>. For convenience, the service also has a Map<String, Team> so that individual teams can be accessed by name. Here is the code segment:

```
private void deserialize() {
    // Deserialize the bytes into a list of teams
    XMLDecoder dec = new XMLDecoder(new ByteArrayInputStream(team_bytes));
    teams = (List<Team>) dec.readObject();

    // Create a map for quick lookups of teams.
    team_map = Collections.synchronizedMap(new HashMap<String, Team>());
    for (Team team : teams) team_map.put(team.getName(), team);
}
```

The RestfulTeams service is published using the by-now-familiar Endpoint publisher, the same publisher used for SOAP-based services under JWS:

```
package ch04.team;

import javax.xml.ws.Endpoint;

class TeamsPublisher {
    public static void main(String[ ] args) {
        int port = 8888;
        String url = "http://localhost:" + port + "/teams";
        System.out.println("Publishing Teams restfully on port " + port);
        Endpoint.publish(url, new RestfulTeams());
    }
}
```

Of the four HTTP verbs that correspond to CRUD operations, only GET has no side effects on the resource, which is the list of classic comedy teams. For now, then, there is no need to serialize a changed List<Team> to the file *teams.ser*.

The JWS runtime dispatches client requests against the RestfulTeams service to the invoke method:

```
public Source invoke(Source request) {
    if (ws_ctx == null) throw new RuntimeException("Injection failed on ws_ctx.");

    // Grab the message context and extract the request verb.
    MessageContext msg_ctx = ws_ctx.getMessageContext();
    String http_verb = (String) msg_ctx.get(MessageContext.HTTP_REQUEST_METHOD);
    http_verb = http_verb.trim().toUpperCase();

    // Act on the verb. For now, only GET requests accepted.
    if (http_verb.equals("GET")) return doGet(msg_ctx);
    else throw new HTTPException(405); // method not allowed
}
```

This method extracts the HTTP request verb from the MessageContext and then invokes a verb-appropriate method such as doGet to handle the request. If the request verb is not GET, then an HTTPException is thrown with the status code 405 to signal *method not allowed*. Table 4-3 shows some of the many HTTP status codes.

Table 4-3. Sample HTTP status codes

HTTP status code	Official reason	Meaning
200	OK	Request OK.
400	Bad request	Request malformed.
403	Forbidden	Request refused.
404	Not found	Resource not found.
405	Method not allowed	Method not supported.
415	Unsupported media type	Content type not recognized.
500	Internal server error	Request processing failed.

In general, status codes in the range of 100–199 are informational; those in the range of 200–299 are success codes; codes in the range of 300–399 are for redirection; those in the range of 400–499 signal client errors; and codes in the range of 500–599 indicate server errors.

There are two types of GET (and, later, DELETE) requests handled in the service. If the GET request comes without a query string, the RestfulTeams service treats this as a request for the entire list of teams and responds with a copy of the XML document in the file *teams.ser*. If the GET request has a query string, this should be in the form ?name=<*team name*>, for instance, ?name=MarxBrothers. In this case, the doGet method gets the named team and encodes this team as an XML document using the XMLEncoder in the method encode_to_stream. Here is the body of the doGet method:

```
if (query_string == null) // get all teams
    // Respond with list of all teams
    return new StreamSource(new ByteArrayInputStream(team_bytes));
else { // get the named team
    String name = get_name_from_qs(query_string);

    // Check if named team exists.
    Team team = team_map.get(name);
    if (team == null) throw new HTTPException(404); // not found

    // Respond with named team.
    ByteArrayInputStream stream = encode_to_stream(team);
    return new StreamSource(stream);
}
```

The StreamSource is a source of bytes that come from the XML document and are made available to the requesting client. On a request for the Marx Brothers, the doGet method returns, as a byte stream, an XML document that begins:

```
<java version="1.6.0_06" class="java.beans.XMLDecoder">
 <object class="ch04.team.Team">
  <void property="name">
   <string>MarxBrothers</string>
  </void>
  <void property="players">
   <object class="java.util.ArrayList">
    <void method="add">
     <object class="ch04.team.Player">
      <void property="name">
       <string>Leonard Marx</string>
      </void>
      <void property="nickname">
       <string>Chico</string>
      ...
```

Language Transparency and RESTful Services

As evidence of language transparency, the first client against the RestfulTeams service is not in Java but rather in Perl. The client sends two GET requests and performs elementary processing on the responses. Here is the initial Perl client:

```perl
#!/usr/bin/perl

use strict;
use LWP;
use XML::XPath;

# Create the user agent.
my $ua = LWP::UserAgent->new;

my $base_uri = 'http://localhost:8888/teams';

# GET teams?name=MarxBrothers
my $request = $base_uri . '?name=MarxBrothers';
send_GET($request);

sub send_GET {
    my ($uri, $qs_flag) = @_;

    # Send the request and get the response.
    my $req = HTTP::Request->new(GET => $uri);
    my $res = $ua->request($req);

    # Check for errors.
    if ($res->is_success) {
        parse_GET($res->content, $qs_flag); # Process raw XML on success
    }
    else {
        print $res->status_line, "\n";      # Print error code on failure
    }
}

# Print raw XML and the elements of interest.
sub parse_GET {
    my ($raw_xml) = @_;
    print "\nThe raw XML response is:\n$raw_xml\n;;;\n";

    # For all teams, extract and print out their names and members
    my $xp = XML::XPath->new(xml => $raw_xml);
    foreach my $node ($xp->find('//object/void/string')->get_nodelist) {
        print $node->string_value, "\n";
    }
}
```

The Perl client issues a GET request against the URI *http://localhost:8888/teams*, which is the endpoint location for the Endpoint-published service. If the request succeeds, the service returns an XML representation of the teams, in this case the XML generated from a call to the XMLEncoder method writeObject. The Perl client prints the raw XML and performs a very simple parse, using an XPath package to get the team names

---

together with the member names and nicknames. In a production environment the XML processing would be more elaborate, but the basic logic of the client would be the same: issue an appropriate request against the service and process the response in some appropriate way. On a sample client run, the output was:

```
The GET request is: http://localhost:8888/teams
The raw XML response is:
<java version="1.6.0_06" class="java.beans.XMLDecoder">
 <object class="java.util.ArrayList">
  <void method="add">
   <object class="ch04.team.Team">
    <void property="name">
     <string>BurnsAndAllen</string>
    </void>
    <void property="players">
     <object class="java.util.ArrayList">
      <void method="add">
       <object class="ch04.team.Player">
        <void property="name">
         <string>George Burns</string>
        </void>
        <void property="nickname">
         <string>George</string>
        </void>
       </object>
      </void>
      <void method="add">
       <object class="ch04.team.Player">
        <void property="name">
         <string>Gracie Allen</string>
        </void>
        <void property="nickname">
         <string>Gracie</string>
        </void>
       </object>
      </void>
     </object>
    </void>
   </object>
  </void>
  ...
</java>
;;;

BurnsAndAllen
George Burns
George
Gracie Allen
Gracie
AbbottAndCostello
William Abbott
Bud
Louis Cristillo
Lou
MarxBrothers
```

```
Leonard Marx
Chico
Julius Marx
Groucho
Adolph Marx
Harpo
```

The output below the semicolons consists of the extracted team names, together with the member names and nicknames.

Here is a Java client against the RestfulTeams service:

```java
import java.util.Arrays;
import java.net.URL;
import java.net.HttpURLConnection;
import java.net.MalformedURLException;
import java.net.URLEncoder;
import java.io.IOException;
import java.io.PrintWriter;
import java.io.BufferedReader;
import java.io.InputStreamReader;
import java.io.ByteArrayInputStream;
import org.xml.sax.helpers.DefaultHandler;
import org.xml.sax.Attributes;
import org.xml.sax.SAXException;
import javax.xml.parsers.SAXParserFactory;
import javax.xml.parsers.SAXParser;
import javax.xml.parsers.ParserConfigurationException;

class TeamsClient {
    private static final String endpoint = "http://localhost:8888/teams";

    public static void main(String[ ] args) {
        new TeamsClient().send_requests();
    }

    private void send_requests() {
        try {
            // GET requests
            HttpURLConnection conn = get_connection(endpoint, "GET");
            conn.connect();
            print_and_parse(conn, true);

            conn = get_connection(endpoint + "?name=MarxBrothers", "GET");
            conn.connect();
            print_and_parse(conn, false);
        }
        catch(IOException e) { System.err.println(e); }
        catch(NullPointerException e) { System.err.println(e); }
    }

    private HttpURLConnection get_connection(String url_string,
                                             String verb) {
        HttpURLConnection conn = null;
        try {
            URL url = new URL(url_string);
```

```
                conn = (HttpURLConnection) url.openConnection();
                conn.setRequestMethod(verb);
        }
        catch(MalformedURLException e) { System.err.println(e); }
        catch(IOException e) { System.err.println(e); }
        return conn;
    }

    private void print_and_parse(HttpURLConnection conn, boolean parse) {
        try {
            String xml = "";
            BufferedReader reader =
                new BufferedReader(new InputStreamReader(conn.getInputStream()));
            String next = null;
            while ((next = reader.readLine()) != null)
                xml += next;
            System.out.println("The raw XML:\n" + xml);

            if (parse) {
                SAXParser parser =SAXParserFactory.newInstance().newSAXParser();
                parser.parse(new ByteArrayInputStream(xml.getBytes()),
                                new SaxParserHandler());
            }
        }
        catch(IOException e) { System.err.println(e); }
        catch(ParserConfigurationException e) { System.err.println(e); }
        catch(SAXException e) { System.err.println(e); }
    }

    static class SaxParserHandler extends DefaultHandler {
        char[ ] buffer = new char[1024];
        int n = 0;

        public void startElement(String uri, String lname,
                                    String qname, Attributes attributes) {
            clear_buffer();
        }

        public void characters(char[ ] data, int start, int length) {
            System.arraycopy(data, start, buffer, 0, length);
            n += length;
        }

        public void endElement(String uri, String lname, String qname) {
            if (Character.isUpperCase(buffer[0]))
                System.out.println(new String(buffer));
            clear_buffer();
        }

        private void clear_buffer() {
            Arrays.fill(buffer, '\0');
            n = 0;
        }
    }
}
```

The Java client issues two GET requests and uses a *SAX* (Simple API for XML) parser to process the returned XML. Java offers an assortment of XML-processing tools and the code examples illustrate several. A SAX parser is stream-based and event-driven— the parser receives a stream of bytes, invoking callbacks (such as the methods named `startElement` and `characters` shown above) to handle specific events, in this case the occurrence of XML start tags and character data in between start and end tags, respectively.

Summary of the RESTful Features

This first restricted example covers some key features of RESTful services but also ignores one such feature. Following is a summary of the example so far:

- In a request, the pairing of an HTTP verb such as GET with a URI such as *http://.../teams* specifies a CRUD operation against a resource; in this example, a request to read available information about comedy teams.

- The service uses HTTP status codes such as 404 (*resource not found*) and 405 (*method not allowed*) to respond to bad requests.

- If the request is a good one, the service responds with an XML representation that captures the state of the requested resource. So far, the service honors only GET requests, but the other CRUD verbs will be added in the forthcoming revision.

- The service does not take advantage of MIME types. A client issues a request for either a named team or a list of all teams but does not indicate a preference for the type of representation returned (for instance, `text/plain` as opposed to `text/xml` or `text/html`). A later example does illustrate typed requests and responses.

- The RESTful service implementation is not constrained in the same way as a SOAP-based service precisely because there is no formal service contract. The implementation is flexible but, of course, likewise ad hoc. This issue will be raised often.

The next section extends the service to handle requests issued with the POST, PUT, and DELETE verbs.

Implementing the Remaining CRUD Operations

The remaining CRUD operations—*create* (POST), *update* (PUT), and *delete* (DELETE)—have side effects, which requires that the `RestfulTeams` service update the in-memory data structures (in this case, the list and the map of teams) and the persistence store (in this case, the local file *teams.ser*). The service follows an eager rather than a lazy strategy for updating *teams.ser*—this file is updated on every successful POST, PUT, and DELETE request. A lazier and more efficient strategy might be followed in a production environment.

The `RestfulTeams` implementation of the `invoke` method changes only slightly to accommodate the new request possibilities. Here is the change:

```
MessageContext msg_ctx = ws_ctx.getMessageContext();
String http_verb = (String) msg_ctx.get(MessageContext.HTTP_REQUEST_METHOD);
http_verb = http_verb.trim().toUpperCase();

// Act on the verb.
if      (http_verb.equals("GET"))     return doGet(msg_ctx);
else if (http_verb.equals("DELETE"))  return doDelete(msg_ctx);
else if (http_verb.equals("POST"))    return doPost(msg_ctx);
else if (http_verb.equals("PUT"))     return doPut(msg_ctx);
else throw new HTTPException(405);     // method not allowed
```

The doPost method expects that the request contains an XML document with information about the new team to be created. Following is a sample:

```
<create_team>
    <name>SmothersBrothers</name>
    <player>
      <name>Thomas</name>
      <nickname>Tom</nickname>
    </player>
    <player>
      <name>Richard</name>
      <nickname>Dickie</nickname>
    </player>
</create_team>
```

Of course, an XML Schema that describes precisely this layout could be distributed to clients. In this example, the doPost does not validate the request document against a schema but rather parses the document to find required information such as the team's name and the players' names. If required information is missing, an HTTP status code of 500 (*internal error*) or 400 (*bad request*) is sent back to the client. Here is the added doPost method:

```
private Source doPost(MessageContext msg_ctx) {
    Map<String, List> request = (Map<String, List>)
      msg_ctx.get(MessageContext.HTTP_REQUEST_HEADERS);

    List<String> cargo = request.get(post_put_key);
    if (cargo == null) throw new HTTPException(400); // bad request

    String xml = "";
    for (String next : cargo) xml += next.trim();
    ByteArrayInputStream xml_stream = new ByteArrayInputStream(xml.getBytes());
    String team_name = null;

    try {
        // Set up the XPath object to search for the XML elements.
        DOMResult dom = new DOMResult();
        Transformer trans = TransformerFactory.newInstance().newTransformer();
        trans.transform(new StreamSource(xml_stream), dom);
        URI ns_URI = new URI("create_team");

        XPathFactory xpf = XPathFactory.newInstance();
        XPath xp = xpf.newXPath();
        xp.setNamespaceContext(new NSResolver("", ns_URI.toString()));
```

```
        team_name = xp.evaluate("/create_team/name", dom.getNode());
        List<Player> team_players = new ArrayList<Player>();
        NodeList players = (NodeList) xp.evaluate("player", dom.getNode(),
                                        XPathConstants.NODESET);

        for (int i = 1; i <= players.getLength(); i++) {
            String name = xp.evaluate("name", dom.getNode());
            String nickname = xp.evaluate("nickname", dom.getNode());
            Player player = new Player(name, nickname);
            team_players.add(player);
        }

        // Add new team to the in-memory map and save List to file.
        Team t = new Team(team_name, team_players);
        team_map.put(team_name, t);
        teams.add(t);
        serialize();
    }
    catch(URISyntaxException e) { throw new HTTPException(500); }
    catch(TransformerConfigurationException e) { throw new HTTPException(500); }
    catch(TransformerException e) { throw new HTTPException(500); }
    catch(XPathExpressionException e) { throw new HTTPException(400); }
    // Send a confirmation to requester.
    return response_to_client("Team " + team_name + " created.");
}
```

Java API for XML Processing

In parsing the request XML document, the `doPost` method in this example uses interfaces and classes from the `javax.xml.transform` package, which are part of *JAX-P* (Java API for XML-Processing). The JAX-P tools were designed to facilitate XML processing, which addresses the needs of a RESTful service. In this example, the two key pieces are the `DOMResult` and the `XPath` object. In the Java `TeamsClient` shown earlier, a SAX parser is used to process the list of comedy teams returned from the `RestfulTeams` service on a successful GET request with no query string. A SAX parser is stream-based and invokes programmer-supplied callbacks to process various parsing events such as the occurrence of an XML start tag. By contrast, a *DOM* (Document Object Model) parser is tree-based in that the parser constructs a tree representation of a well-formed XML document. The programmer then can use a standard API, for example, to search the tree for desired elements. JAX-P uses the *XSLT* (eXtensible Stylesheet Language Transformations) verb *transform* to describe the process of transforming an XML *source* (for instance, the request bytes from a client) into an XML *result* (for instance, a DOM tree). Here is the statement in `doPost` that does just this:

```
    trans.transform(new StreamSource(xml_stream), dom);
```

The `xml_stream` refers to the bytes from the client in a `ByteArrayInputStream`, and `dom` refers to a `DOMResult`. A DOM tree can be processed in various ways. In this case, an `XPath` object is used to search for relatively simple patterns. For instance, the statement:

```
NodeList players = (NodeList) xp.evaluate("player", dom.getNode(),
                                          XPathConstants.NODESET);
```

gets a list of elements tagged with **player** from the DOM tree. The statements:

```
String name = xp.evaluate("name", dom.getNode());
String nickname = xp.evaluate("nickname", dom.getNode());
```

then extract the player's name and nickname from the DOM tree.

The **doPost** method respects the HTTP verb from which the method gets its name. After the name of the new team has been extracted from the request XML document, a check is made:

```
team_name = xp.evaluate("/create_team/name", dom.getNode());
if (team_map.containsKey(team_name)) throw new HTTPException(400); // bad request
```

to determine whether a team with that name already exists. Because a POST request signals a *create* operation, an already existing team cannot be created but instead must be *updated* through a PUT request.

Once the needed information about the new team has been extracted from the request XML document, the data structures **Map<String, Team>** and **List<Team>** are updated to reflect a successful *create* operation. The list of teams is serialized to the persistence file.

The two remaining CRUD operations, *update* and *delete*, are implemented as the methods **doPut** and **doDelete**, respectively. The **RestfulTeams** service requires that a DELETE request have a query string to identify a particular team; the deletion of all teams at once is not allowed. For now, a PUT request can update only a team's name, although this easily could be expanded to allow updates to the team's members and their names or nicknames. Here are the implementations of **doPut** and **doDelete**:

```
private Source doDelete(MessageContext msg_ctx) {
    String query_string = (String) msg_ctx.get(MessageContext.QUERY_STRING);

    // Disallow the deletion of all teams at once.
    if (query_string == null) throw new HTTPException(403); // illegal operation
    else {
        String name = get_value_from_qs("name", query_string);
        if (!team_map.containsKey(name)) throw new HTTPException(404);

        // Remove team from Map and List, serialize to file.
        Team team = team_map.get(name);
        teams.remove(team);
        team_map.remove(name);
        serialize();

        // Send response.
        return response_to_client(name + " deleted.");
    }
}

private Source doPut(MessageContext msg_ctx) {
    // Parse the query string.
    String query_string = (String) msg_ctx.get(MessageContext.QUERY_STRING);
```

```
        String name = null;
        String new_name = null;

        // Get all teams.
        if (query_string == null) throw new HTTPException(403); // illegal operation
        // Get a named team.
        else {
            // Split query string into name= and new_name= sections
            String[ ] parts = query_string.split("&");
            if (parts[0] == null || parts[1] == null) throw new HTTPException(403);

            name = get_value_from_qs("name", parts[0]);
            new_name = get_value_from_qs("new_name", parts[1]);
            if (name == null || new_name == null) throw new HTTPException(403);

            Team team = team_map.get(name);
            if (team == null) throw new HTTPException(404);
            team.setName(new_name);
            team_map.put(new_name, team);
            serialize();
        }

        // Send a confirmation to requester.
        return response_to_client("Team " + name + " changed to " + new_name);
    }
```

Each of the **do** methods has a similar style, and the application logic has been kept as
simple as possible to focus attention on RESTful character of the service. Here, for
reference, is the all of the source code for the service:

```
package ch04.team;

import javax.xml.ws.Provider;
import javax.xml.transform.Source;
import javax.xml.transform.stream.StreamSource;
import javax.annotation.Resource;
import javax.xml.ws.BindingType;
import javax.xml.ws.WebServiceContext;
import javax.xml.ws.handler.MessageContext;
import javax.xml.ws.http.HTTPException;
import javax.xml.ws.WebServiceProvider;
import javax.xml.ws.ServiceMode;
import javax.xml.ws.http.HTTPBinding;
import java.io.ByteArrayInputStream;
import java.io.ByteArrayOutputStream;
import java.util.Collections;
import java.util.Map;
import java.util.HashMap;
import java.util.List;
import java.util.ArrayList;
import java.io.IOException;
import java.io.File;
import java.io.FileInputStream;
import java.io.FileOutputStream;
import java.io.BufferedOutputStream;
```

```
import java.beans.XMLEncoder;
import java.beans.XMLDecoder;
import javax.xml.transform.TransformerFactory;
import javax.xml.transform.Transformer;
import javax.xml.transform.dom.DOMResult;
import javax.xml.transform.TransformerException;
import javax.xml.transform.TransformerConfigurationException;
import javax.xml.xpath.XPathFactory;
import javax.xml.xpath.XPath;
import javax.xml.xpath.XPathConstants;
import javax.xml.xpath.XPathExpressionException;
import java.net.URI;
import java.net.URISyntaxException;
import org.w3c.dom.NodeList;

// The class below is a WebServiceProvider rather than
// the more usual SOAP-based WebService. As a result, the
// service implements the generic Provider interface rather
// than a customized SEI with designated @WebMethods.
@WebServiceProvider

// There are two ServiceModes: PAYLOAD, the default, signals that the service
// wants access only to the underlying message payload (e.g., the
// body of an HTTP POST request); MESSAGE signals that the service wants
// access to entire message (e.g., the HTTP headers and body). In this
// case, the MESSAGE mode lets us check on the request verb.
@ServiceMode(value = javax.xml.ws.Service.Mode.MESSAGE)

// The HTTP_BINDING as opposed, for instance, to a SOAP binding.
@BindingType(value = HTTPBinding.HTTP_BINDING)

// The generic, low-level Provider interface is an alternative
// to the SEI (service endpoint interface) of a SOAP-based
// web service. A Source is a source of the bytes. The invoke
// method expects a source and returns one.
public class RestfulTeams implements Provider<Source> {
    @Resource
    protected WebServiceContext ws_ctx;

    private Map<String, Team> team_map; // for easy lookups
    private List<Team> teams;           // serialized/deserialized
    private byte[ ] team_bytes;         // from the persistence file

    private static final String file_name = "teams.ser";
    private static final String post_put_key = "Cargo";

    public RestfulTeams() {
        read_teams_from_file();
        deserialize();
    }

    // Implementation of the Provider interface method: this
    // method handles incoming requests and generates the
    // outgoing response.
    public Source invoke(Source request) {
```

```
    if (ws_ctx == null)
        throw new RuntimeException("Injection failed on ws_ctx.");

    if (request == null) System.out.println("null request");
    else System.out.println("non-null request");

    // Grab the message context and extract the request verb.
    MessageContext msg_ctx = ws_ctx.getMessageContext();
    String http_verb = (String)
        msg_ctx.get(MessageContext.HTTP_REQUEST_METHOD);
    http_verb = http_verb.trim().toUpperCase();

    // Act on the verb.
    if      (http_verb.equals("GET"))    return doGet(msg_ctx);
    else if (http_verb.equals("DELETE")) return doDelete(msg_ctx);
    else if (http_verb.equals("POST"))   return doPost(msg_ctx);
    else if (http_verb.equals("PUT"))    return doPut(msg_ctx);
    else throw new HTTPException(405);    // bad verb exception
}

private Source doGet(MessageContext msg_ctx) {
    // Parse the query string.
    String query_string = (String)
        msg_ctx.get(MessageContext.QUERY_STRING);

    // Get all teams.
    if (query_string == null)
        return new StreamSource(new ByteArrayInputStream(team_bytes));
    // Get a named team.
    else {
        String name = get_value_from_qs("name", query_string);

        // Check if named team exists.
        Team team = team_map.get(name);
        if (team == null) throw new HTTPException(404); // not found

        // Otherwise, generate XML and return.
        ByteArrayInputStream stream = encode_to_stream(team);
        return new StreamSource(stream);
    }
}

private Source doPost(MessageContext msg_ctx) {
    Map<String, List> request = (Map<String, List>)
        msg_ctx.get(MessageContext.HTTP_REQUEST_HEADERS);

    List<String> cargo = request.get(post_put_key);
    if (cargo == null) throw new HTTPException(400); // bad request

    String xml = "";
    for (String next : cargo) xml += next.trim();
    ByteArrayInputStream xml_stream = new ByteArrayInputStream(xml.getBytes());
    String team_name = null;
```

```java
try {
    // Set up the XPath object to search for the XML elements.
    DOMResult dom = new DOMResult();
    Transformer trans =
        TransformerFactory.newInstance().newTransformer();
    trans.transform(new StreamSource(xml_stream), dom);
    URI ns_URI = new URI("create_team");

    XPathFactory xpf = XPathFactory.newInstance();
    XPath xp = xpf.newXPath();
    xp.setNamespaceContext(new NSResolver("", ns_URI.toString()));

    team_name = xp.evaluate("/create_team/name", dom.getNode());

    if (team_map.containsKey(team_name))
        throw new HTTPException(400); // bad request

    List<Player> team_players = new ArrayList<Player>();

    NodeList players = (NodeList)
        xp.evaluate("player",
                    dom.getNode(),
                    XPathConstants.NODESET);

    for (int i = 1; i <= players.getLength(); i++) {
        String name = xp.evaluate("name", dom.getNode());
        String nickname = xp.evaluate("nickname", dom.getNode());
        Player player = new Player(name, nickname);
        team_players.add(player);
    }
    // Add new team to the in-memory map and save List to file.
    Team t = new Team(team_name, team_players);
    team_map.put(team_name, t);
    teams.add(t);
    serialize();
}
catch(URISyntaxException e) {
    throw new HTTPException(500);   // internal server error
}
catch(TransformerConfigurationException e) {
    throw new HTTPException(500);   // internal server error
}
catch(TransformerException e) {
    throw new HTTPException(500);   // internal server error
}
catch(XPathExpressionException e) {
    throw new HTTPException(400);   // bad request
}

// Send a confirmation to requester.
return response_to_client("Team " + team_name + " created.");
}
```

```
private Source doPut(MessageContext msg_ctx) {
    // Parse the query string.
    String query_string = (String) msg_ctx.get(MessageContext.QUERY_STRING);
    String name = null;
    String new_name = null;

    // Get all teams.
    if (query_string == null)
        throw new HTTPException(403); // illegal operation
    // Get a named team.
    else {
        // Split query string into name= and new_name= sections
        String[ ] parts = query_string.split("&");
        if (parts[0] == null || parts[1] == null)
            throw new HTTPException(403);

        name = get_value_from_qs("name", parts[0]);
        new_name = get_value_from_qs("new_name", parts[1]);
        if (name == null || new_name == null)
            throw new HTTPException(403);

        Team team = team_map.get(name);
        if (team == null) throw new HTTPException(404);
        team.setName(new_name);
        team_map.put(new_name, team);
        serialize();
    }

    // Send a confirmation to requester.
    return response_to_client("Team " + name + " changed to " + new_name);
}

private Source doDelete(MessageContext msg_ctx) {
    String query_string = (String)
        msg_ctx.get(MessageContext.QUERY_STRING);

    // Disallow the deletion of all teams at once.
    if (query_string == null)
        throw new HTTPException(403);      // illegal operation
    else {
        String name = get_value_from_qs("name", query_string);
        if (!team_map.containsKey(name))
            throw new HTTPException(404); // not found

        // Remove team from Map and List, serialize to file.
        Team team = team_map.get(name);
        teams.remove(team);
        team_map.remove(name);
        serialize();

        // Send response.
        return response_to_client(name + " deleted.");
    }
}
```

```java
private StreamSource response_to_client(String msg) {
    HttpResponse response = new HttpResponse();
    response.setResponse(msg);
    ByteArrayInputStream stream = encode_to_stream(response);
    return new StreamSource(stream);
}

private ByteArrayInputStream encode_to_stream(Object obj) {
    // Serialize object to XML and return
    ByteArrayOutputStream stream = new ByteArrayOutputStream();
    XMLEncoder enc = new XMLEncoder(stream);
    enc.writeObject(obj);
    enc.close();
    return new ByteArrayInputStream(stream.toByteArray());
}

private String get_value_from_qs(String key, String qs) {
    String[ ] parts = qs.split("=");

    // Check if query string has form: name=<team name>
    if (!parts[0].equalsIgnoreCase(key))
        throw new HTTPException(400); // bad request
    return parts[1].trim();
}

private void read_teams_from_file() {
    try {
        String cwd = System.getProperty ("user.dir");
        String sep = System.getProperty ("file.separator");
        String path = get_file_path();
        int len = (int) new File(path).length();
        team_bytes = new byte[len];
        new FileInputStream(path).read(team_bytes);
    }
    catch(IOException e) { System.err.println(e); }
}

private void deserialize() {
    // Deserialize the bytes into a list of teams
    XMLDecoder dec =
        new XMLDecoder(new ByteArrayInputStream(team_bytes));
    teams = (List<Team>) dec.readObject();

    // Create a map for quick lookups of teams.
    team_map = Collections.synchronizedMap(new HashMap<String, Team>());
    for (Team team : teams)
        team_map.put(team.getName(), team);
}

private void serialize() {
    try {
        String path = get_file_path();
        BufferedOutputStream out =
            new BufferedOutputStream(new FileOutputStream(path));
```

```
            XMLEncoder enc = new XMLEncoder(out);
            enc.writeObject(teams);
            enc.close();
            out.close();
        }
        catch(IOException e) { System.err.println(e); }
    }

    private String get_file_path() {
        String cwd = System.getProperty ("user.dir");
        String sep = System.getProperty ("file.separator");
        return cwd + sep + "ch04" + sep + "team" + sep + file_name;
    }
}
```

The revised Perl client shown below tests the service by generating a series of requests.
Here is the complete Perl client:

```perl
#!/usr/bin/perl

use strict;
use LWP;
use XML::XPath;
use Encode;
use constant true   => 1;
use constant false  => 0;

# Create the user agent.
my $ua = LWP::UserAgent->new;

my $base_uri = 'http://localhost:8888/teams';

# GET teams
send_GET($base_uri, false); # false means no query string

# GET teams?name=MarxBrothers
send_GET($base_uri . '?name=MarxBrothers', true);

$base_uri = $base_uri;
send_POST($base_uri);

# Check that POST worked
send_GET($base_uri . '?name=SmothersBrothers', true);
send_DELETE($base_uri . '?name=SmothersBrothers');

# Recreate the Smothers Brothers as a check.
send_POST($base_uri);

# Change name and check.
send_PUT($base_uri . '?name=SmothersBrothers&new_name=SmuthersBrothers');
send_GET($base_uri . '?name=SmuthersBrothers', true);

sub send_GET {
    my ($uri, $qs_flag) = @_;
```

```
    # Send the request and get the response.
    my $req = HTTP::Request->new(GET => $uri);
    my $res = $ua->request($req);

    # Check for errors.
    if ($res->is_success) {
        parse_GET($res->content, $qs_flag); # Process raw XML on success
    }
    else {
        print $res->status_line, "\n";      # Print error code on failure
    }
}

sub send_POST {
    my ($uri) = @_;

    my $xml = <<EOS;
      <create_team>
        <name>SmothersBrothers</name>
        <player>
          <name>Thomas</name>
          <nickname>Tom</nickname>
        </player>
        <player>
          <name>Richard</name>
          <nickname>Dickie</nickname>
        </player>
      </create_team>
EOS
    # Send request and capture response.
    my $bytes = encode('iso-8859-1', $xml); # encoding is Latin-1
    my $req = HTTP::Request->new(POST => $uri, ['Cargo' => $bytes]);
    my $res = $ua->request($req);

    # Check for errors.
    if ($res->is_success) {
        parse_SIMPLE("POST", $res->content); # Process raw XML on success
    }
    else {
        print $res->status_line, "\n";       # Print error code on failure
    }
}

sub send_DELETE {
    my $uri = shift;

    # Send the request and get the response.
    my $req = HTTP::Request->new(DELETE => $uri);
    my $res = $ua->request($req);

    # Check for errors.
    if ($res->is_success) {
        parse_SIMPLE("DELETE", $res->content);   # Process raw XML on success
    }
```

```
        else {
            print $res->status_line, "\n"; # Print error code on failure
        }
    }

    sub send_PUT {
        my $uri = shift;

        # Send the request and get the response.
        my $req = HTTP::Request->new(PUT => $uri);
        my $res = $ua->request($req);

        # Check for errors.
        if ($res->is_success) {
            parse_SIMPLE("PUT", $res->content);    # Process raw XML on success
        }
        else {
            print $res->status_line, "\n"; # Print error code on failure
        }
    }

    sub parse_SIMPLE {
        my $verb = shift;
        my $raw_xml = shift;
        print "\nResponse on $verb: \n$raw_xml;;;\n";
    }

    sub parse_GET {
        my ($raw_xml) = @_;
        print "\nThe raw XML response is:\n$raw_xml\n;;;\n";

        # For all teams, extract and print out their names and members
        my $xp = XML::XPath->new(xml => $raw_xml);
        foreach my $node ($xp->find('//object/void/string')->get_nodelist) {
            print $node->string_value, "\n";
        }
    }
```

The Provider and Dispatch Twins

In the RestfulTeams service, the clients send request information to the service through
the HTTP start line (for instance, in a GET request) and optionally through an inserted
HTTP header (for instance, in a POST request). Recall that GET and DELETE requests
result in HTTP messages that have no body, whereas POST and PUT requests result
in HTTP messages with bodies. Clients of the RestfulTeams service do not use the HTTP
body at all. Even in a POST or PUT request, information about the new team to create
or the existing team to update is contained in the HTTP header rather than in the body.

The approach in the RestfulTeams service illustrates the flexibility of REST-style services. The revision in this section shows how the HTTP body can be used in a POST
request by introducing the Dispatch interface, which is the client-side twin of the server-
side Provider interface. The RestfulTeams service already illustrates that a Provider on

the service side can be used without a `Dispatch` on the client side; and a later example shows how a `Dispatch` can be used on the client side regardless of how the RESTful service is implemented. Nonetheless, the `Provider` and `Dispatch` interfaces are a natural pair.

A RESTful `Provider` implements the method:

```
public Source invoke(Source request)
```

and a `Dispatch` object, sometimes described as a *dynamic service proxy*, provides an implementation of this method on the client side. Recall that a `Source` is a source of an XML document suitable as input to a `Transform`, which then generates a `Result` that is typically an XML document as well. The `Dispatch` to `Provider` relationship supports a natural exchange of XML documents between client and service:

- The client invokes the `Dispatch` method `invoke`, with an XML document as the `Source` argument. If the request does not require an XML document, then the `Source` argument can be `null`.

- The service-side runtime dispatches the client request to the `Provider` method `invoke` whose `Source` argument corresponds to the client-side `Source`.

- The service transforms the `Source` into some appropriate `Result` (for instance, a DOM tree), processes this `Result` in an application-appropriate way, and returns an XML source to the client. If no response is needed, `null` can be returned.

- The `Dispatch` method `invoke` returns a `Source`, sent from the service, that the client then transforms into an appropriate `Result` and processes as needed.

The fact that the `Provider` method `invoke` and the `Dispatch` method `invoke` have the same signature underscores the natural fit between them.

A Provider/Dispatch Example

The `RabbitCounterProvider` is a RESTful service that revises the SOAP-based version of Chapter 3. The RESTful revision honors POST, GET, and DELETE requests from clients. A POST request, as a CRUD *create* operation, creates a list of Fibonacci numbers that the service caches for subsequent *read* or *delete* operations. The `doPost` method responds to a POST request and the method expects a `Source` argument, which is the source of an XML document such as:

```
<fib:request xmlns:fib = 'urn:fib'>[1, 2, 3, 4]</fib:request>
```

The XML document is thus a list of integers whose Fibonacci values are to be computed. The `doGet` and `doDelete` methods handle GET and PUT requests, respectively, neither of which has an HTTP body; hence, the `doGet` and `doDelete` methods do not have a `Source` parameter. All three methods return a `Source` value, which is the source of an XML confirmation. For example, `doPost` returns a confirmation XML document such as:

```
<fib:response xmlns:fib = 'urn:fib'>POSTed[1, 1, 2, 3]</fib:response>
```

The other two methods return operation-specific confirmations.

Here is the source code for the RabbitCounterProvider:

```
package ch04.dispatch;

import java.util.Collections;
import java.util.List;
import java.util.ArrayList;
import java.util.Map;
import java.util.HashMap;
import java.util.Collection;
import javax.xml.ws.Provider;
import javax.xml.transform.Source;
import javax.xml.transform.stream.StreamSource;
import javax.annotation.Resource;
import javax.xml.ws.BindingType;
import javax.xml.ws.WebServiceContext;
import javax.xml.ws.handler.MessageContext;
import javax.xml.ws.http.HTTPException;
import javax.xml.ws.WebServiceProvider;
import javax.xml.ws.http.HTTPBinding;
import java.io.ByteArrayInputStream;
import javax.xml.transform.TransformerFactory;
import javax.xml.transform.Transformer;
import javax.xml.transform.dom.DOMResult;
import javax.xml.transform.TransformerException;
import javax.xml.transform.TransformerConfigurationException;
import javax.xml.xpath.XPathFactory;
import javax.xml.xpath.XPath;
import javax.xml.xpath.XPathConstants;
import javax.xml.xpath.XPathExpressionException;

// The RabbitCounter service implemented as REST style rather than SOAP based.
@WebServiceProvider
@BindingType(value = HTTPBinding.HTTP_BINDING)

public class RabbitCounterProvider implements Provider<Source> {
    @Resource
    protected WebServiceContext ws_ctx;

    // stores previously computed values
    private Map<Integer, Integer> cache =
        Collections.synchronizedMap(new HashMap<Integer, Integer>());

    private final String xml_start = "<fib:response xmlns:fib = 'urn:fib'>";
    private final String xml_stop = "</fib:response>";
    private final String uri = "urn:fib";

    public Source invoke(Source request) {
        // Filter on the HTTP request verb
        if (ws_ctx == null) throw new RuntimeException("DI failed on ws_ctx.");
```

```
        // Grab the message context and extract the request verb.
        MessageContext msg_ctx = ws_ctx.getMessageContext();
        String http_verb = (String) msg_ctx.get(MessageContext.HTTP_REQUEST_METHOD);
        http_verb = http_verb.trim().toUpperCase();

        // Act on the verb.
        if      (http_verb.equals("GET"))    return doGet();
        else if (http_verb.equals("DELETE")) return doDelete();
        else if (http_verb.equals("POST"))   return doPost(request);
        else throw new HTTPException(405);    // bad verb exception
    }

    private Source doPost(Source request) {
        if (request == null) throw new HTTPException(400); // bad request

        String nums = extract_request(request);
        // Extract the integers from a string such as: "[1, 2, 3]"
        nums = nums.replace('[', '\0');
        nums = nums.replace(']', '\0');
        String[ ] parts = nums.split(",");
        List<Integer> list = new ArrayList<Integer>();
        for (String next : parts) {
            int n = Integer.parseInt(next.trim());
            cache.put(n, countRabbits(n));
            list.add(cache.get(n));
        }
        String xml = xml_start + "POSTed: " + list.toString() + xml_stop;
        return make_stream_source(xml);
    }

    private Source doGet() {
        Collection<Integer> list = cache.values();
        String xml = xml_start + "GET: " + list.toString() + xml_stop;
        return make_stream_source(xml);
    }

    private Source doDelete() {
        cache.clear();
        String xml = xml_start + "DELETE: Map cleared." + xml_stop;
        return make_stream_source(xml);
    }

    private String extract_request(Source request) {
        String request_string = null;
        try {
            DOMResult dom_result = new DOMResult();
            Transformer trans = TransformerFactory.newInstance().newTransformer();
            trans.transform(request, dom_result);

            XPathFactory xpf = XPathFactory.newInstance();
            XPath xp = xpf.newXPath();
            xp.setNamespaceContext(new NSResolver("fib", uri));
            request_string = xp.evaluate("/fib:request", dom_result.getNode());
        }
```

```
            catch(TransformerConfigurationException e) { System.err.println(e); }
            catch(TransformerException e) { System.err.println(e); }
            catch(XPathExpressionException e) { System.err.println(e); }

            return request_string;
        }

        private StreamSource make_stream_source(String msg) {
            System.out.println(msg);
            ByteArrayInputStream stream = new ByteArrayInputStream(msg.getBytes());
            return new StreamSource(stream);
        }

        private int countRabbits(int n) {
            if (n < 0) throw new HTTPException(403); // forbidden

            // Easy cases.
            if (n < 2) return n;

            // Return cached values if present.
            if (cache.containsKey(n)) return cache.get(n);
            if (cache.containsKey(n - 1) && cache.containsKey(n - 2)) {
              cache.put(n, cache.get(n - 1) + cache.get(n - 2));
              return cache.get(n);
            }

            // Otherwise, compute from scratch, cache, and return.
            int fib = 1, prev = 0;
            for (int i = 2; i <= n; i++) {
                int temp = fib;
                fib += prev;
                prev = temp;
            }
            cache.put(n, fib);
            return fib;
        }
    }
```

The code segment:

```
    XPathFactory xpf = XPathFactory.newInstance();
    XPath xp = xpf.newXPath();
    xp.setNamespaceContext(new NSResolver("fib", uri));
    request_string = xp.evaluate("/fib:request", dom_result.getNode());
```

deserves a closer look because the NSResolver also is used in the RestfulTeams service.
The call to xp.evaluate, shown in bold above, takes two arguments: an XPath pattern,
in this case /fib:request, and the DOMResult node that contains the desired string data
between the start tag <fib:request> and the corresponding end tag </fib:request>.
The fib in fib:request is a proxy or alias for a namespace URI, in this case urn:fib.
The entire start tag in the request XML document is:

```
    <fib:request xmlns:fib = 'urn:fib'>
```

The NSResolver class (NS is short for namespace) provides mappings from fib to urn:fib and vice-versa. Here is the code:

```
package ch04.dispatch;

import java.util.Collections;
import java.util.Map;
import java.util.HashMap;
import java.util.Iterator;
import javax.xml.namespace.NamespaceContext;

public class NSResolver implements NamespaceContext {
    private Map<String, String> prefix2uri;
    private Map<String, String> uri2prefix;
    public NSResolver() {
        prefix2uri =
            Collections.synchronizedMap(new HashMap<String, String>());
        uri2prefix =
            Collections.synchronizedMap(new HashMap<String, String>());
    }


    public NSResolver(String prefix, String uri) {
        this();
        prefix2uri.put(prefix, uri);
        uri2prefix.put(uri, prefix);
    }

    public String getNamespaceURI(String prefix) { return prefix2uri.get(prefix); }
    public String getPrefix(String uri) { return uri2prefix.get(uri); }
    public Iterator getPrefixes(String uri) { return uri2prefix.keySet().iterator(); }
}
```

The NSResolver provides the namespace context for the XPath searches; that is, the resolver binds together a namespace URI and its proxies or aliases. For the application to work correctly, a client and the service must use the same namespace URI; in this case the structurally simple URI urn:fib.

More on the Dispatch Interface

The Dispatch-based client of the RESTful RabbitCounterProvider service has features reminiscent of a client for a SOAP-based service. The client creates identifying QName instances for a service and a port, creates a service object and adds a port, and then creates a Dispatch proxy associated with the port. Here is the code segment:

```
QName service_name = new QName("rcService", ns_URI.toString()); // uri is urn:fib
QName port = new QName("rcPort", ns_URI.toString());
String endpoint = "http://localhost:9876/fib";
// Now create a service proxy or dispatcher.
Service service = Service.create(service_name);
service.addPort(port, HTTPBinding.HTTP_BINDING, endpoint);
Dispatch<Source> dispatch =
    service.createDispatch(port, Source.class, Service.Mode.PAYLOAD);
```

This client-side `dispatch` object can dispatch XML documents as requests to the service as XML `Source` instances. A document is sent to the service through an invocation of the `invoke` method. Here are two code segments. In the first, an XML document is prepared as the body of a POST request:

```
String xml_start = "<fib:request xmlns:fib = 'urn:fib'>";
String xml_end = "</fib:request>";
List<Integer> nums = new ArrayList<Integer>();
for (int i = 0; i < 12; i++) nums.add(i + 1);
String xml = xml_start + nums.toString() + xml_end;
```

In the second, the request XML document is wrapped in `Source` and then sent to the service through an invocation of `invoke`:

```
StreamSource source = null;
if (data != null) source = make_stream_source(data.toString()); // data = XML doc
Source result = dispatch.invoke(source);
display_result(result, uri); // do an XPath search of the resturned XML
```

The GET and DELETE operations do not require XML documents; hence, the `Source` argument to `invoke` is `null` in both cases. Here is a client-side trace of the requests sent to the service and the responses received in return:

```
Request: <fib:request xmlns:fib = 'urn:fib'>
              [1, 2, 3, 4, 5, 6, 7, 8, 9, 10, 11, 12]
            </fib:request>
POSTed: [1, 1, 2, 3, 5, 8, 13, 21, 34, 55, 89, 144]

Request: null
GET: [1, 1, 2, 3, 5, 8, 13, 21, 34, 55, 89, 144]

Request: null
DELETE: Map cleared.

Request: null
GET: [ ]

Request: <fib:request xmlns:fib = 'urn:fib'>
              [1, 2, 3, 4, 5, 6, 7, 8, 9, 10, 11,...,20, 21, 22, 23, 24]
            </fib:request>
POSTed: [1, 1, 2, 3, 5, 8, 13, 21, 34, 55, 89,..., 10946, 17711, 28657, 46368]

Request: null
GET: [1, 1, 2, 3, 5, 8, 13, 21, 34, 55, 89,..., 10946, 6765, 28657, 17711, 46368]
```

Finally, here is the source code for the entire `DispatchClient`:

```
import java.net.URI;
import java.net.URISyntaxException;
import java.io.ByteArrayInputStream;
import java.util.Map;
import java.util.List;
import java.util.ArrayList;
import javax.xml.namespace.QName;
import javax.xml.ws.Service;
import javax.xml.ws.Dispatch;
```

```
import javax.xml.ws.http.HTTPBinding;
import javax.xml.transform.stream.StreamSource;
import javax.xml.transform.Source;
import javax.xml.transform.TransformerFactory;
import javax.xml.transform.Transformer;
import javax.xml.transform.dom.DOMResult;
import javax.xml.transform.TransformerConfigurationException;
import javax.xml.transform.TransformerException;
import javax.xml.xpath.XPathFactory;
import javax.xml.xpath.XPath;
import javax.xml.xpath.XPathConstants;
import javax.xml.xpath.XPathExpressionException;
import javax.xml.ws.handler.MessageContext;
import org.w3c.dom.NodeList;
import ch04.dispatch.NSResolver;

class DispatchClient {
    public static void main(String[ ] args) throws Exception {
        new DispatchClient().setup_and_test();
    }

    private void setup_and_test() {
        // Create identifying names for service and port.
        URI ns_URI = null;
        try {
            ns_URI = new URI("urn:fib");
        }
        catch(URISyntaxException e) { System.err.println(e); }

        QName service_name = new QName("rcService", ns_URI.toString());
        QName port = new QName("rcPort", ns_URI.toString());
        String endpoint = "http://localhost:9876/fib";

        // Now create a service proxy or dispatcher.
        Service service = Service.create(service_name);
        service.addPort(port, HTTPBinding.HTTP_BINDING, endpoint);
        Dispatch<Source> dispatch =
            service.createDispatch(port, Source.class, Service.Mode.PAYLOAD);

        // Send some requests.
        String xml_start = "<fib:request xmlns:fib = 'urn:fib'>";
        String xml_end = "</fib:request>";

        // To begin, a POST to create some Fibonacci numbers.
        List<Integer> nums = new ArrayList<Integer>();
        for (int i = 0; i < 12; i++) nums.add(i + 1);
        String xml = xml_start + nums.toString() + xml_end;
        invoke(dispatch, "POST", ns_URI.toString(), xml);

        // GET request to test whether the POST worked.
        invoke(dispatch, "GET", ns_URI.toString(), null);

        // DELETE request to remove the list
        invoke(dispatch, "DELETE", ns_URI.toString(), null);
```

```
        // GET to test whether the DELETE worked.
        invoke(dispatch, "GET", ns_URI.toString(), null);

        // POST to repopulate and a final GET to confirm
        nums = new ArrayList<Integer>();
        for (int i = 0; i < 24; i++) nums.add(i + 1);
        xml = xml_start + nums.toString() + xml_end;
        invoke(dispatch, "POST", ns_URI.toString(), xml);
        invoke(dispatch, "GET", ns_URI.toString(), null);
    }

    private void invoke(Dispatch<Source> dispatch,
                        String verb,
                        String uri,
                        Object data) {
        Map<String, Object> request_context = dispatch.getRequestContext();
        request_context.put(MessageContext.HTTP_REQUEST_METHOD, verb);

        System.out.println("Request: " + data);

        // Invoke
        StreamSource source = null;
        if (data != null) source = make_stream_source(data.toString());
        Source result = dispatch.invoke(source);
        display_result(result, uri);
    }

    private void display_result(Source result, String uri) {
        DOMResult dom_result = new DOMResult();
        try {
            Transformer trans = TransformerFactory.newInstance().newTransformer();
            trans.transform(result, dom_result);

            XPathFactory xpf = XPathFactory.newInstance();
            XPath xp = xpf.newXPath();
            xp.setNamespaceContext(new NSResolver("fib", uri));
            String result_string =
                xp.evaluate("/fib:response", dom_result.getNode());
            System.out.println(result_string);
        }
        catch(TransformerConfigurationException e) { System.err.println(e); }
        catch(TransformerException e) { System.err.println(e); }
        catch(XPathExpressionException e) { System.err.println(e); }
    }

    private StreamSource make_stream_source(String msg) {
        ByteArrayInputStream stream = new ByteArrayInputStream(msg.getBytes());
        return new StreamSource(stream);
    }
}
```

A Dispatch Client Against a SOAP-based Service

The Dispatch client is flexible in that it may be used to issue requests against any service, REST-style or SOAP-based. This section illustrates how a SOAP-based service can be treated as if it were REST style. This use of Dispatch underscores that SOAP-based web services delivered over HTTP, as most are, represent a special case of REST-style services. What the SOAP libraries spare the programmer is the need to process XML directly on either the service or the client side, with handlers as the exception to this rule.

The DispatchClientTS application uses a Dispatch proxy to submit a request against the SOAP-based TimeServer service of Chapter 1. The TimeServer supports two operations: one supplies the current time as a human-readable string, whereas the other supplies the time as the elapsed milliseconds from the Unix epoch. Here is the source code for DispatchClientTS:

```
import java.util.Map;
import java.net.URI;
import java.net.URISyntaxException;
import java.io.ByteArrayInputStream;
import javax.xml.namespace.QName;
import javax.xml.ws.Service;
import javax.xml.ws.Dispatch;
import javax.xml.ws.http.HTTPBinding;
import javax.xml.transform.stream.StreamSource;
import javax.xml.transform.Source;
import javax.xml.transform.TransformerFactory;
import javax.xml.transform.Transformer;
import javax.xml.transform.dom.DOMResult;
import javax.xml.transform.TransformerConfigurationException;
import javax.xml.transform.TransformerException;
import javax.xml.xpath.XPathFactory;
import javax.xml.xpath.XPath;
import javax.xml.xpath.XPathConstants;
import javax.xml.xpath.XPathExpressionException;
import javax.xml.ws.handler.MessageContext;
import ch04.dispatch.NSResolver;

// Dispatch client against the SOAP-based TimeServer service
class DispatchClientTS {
    public static void main(String[ ] args) throws Exception {
        new DispatchClientTS().send_and_receive_SOAP();
    }

    private void send_and_receive_SOAP() {
        // Create identifying names for service and port.
        URI ns_URI = null;
        try {
            ns_URI = new URI("http://ts.ch01/");      // from WSDL
        }
        catch(URISyntaxException e) { System.err.println(e); }
```

```java
        QName service_name = new QName("tns", ns_URI.toString());
        QName port = new QName("tsPort", ns_URI.toString());
        String endpoint = "http://localhost:9876/ts"; // from WSDL
        // Now create a service proxy or dispatcher.
        Service service = Service.create(service_name);
        service.addPort(port, HTTPBinding.HTTP_BINDING, endpoint);
        Dispatch<Source> dispatch =
            service.createDispatch(port, Source.class, Service.Mode.PAYLOAD);
        // Send a request.
        String soap_request =
            "<?xml version='1.0' encoding='UTF-8'?> " +
            "<soap:Envelope " +
              "soap:encodingStyle='http://schemas.xmlsoap.org/soap/encoding/' " +
              "xmlns:soap='http://schemas.xmlsoap.org/soap/envelope/' " +
              "xmlns:soapenc='http://schemas.xmlsoap.org/soap/encoding/' " +
              "xmlns:xsi='http://www.w3.org/2001/XMLSchema-instance' " +
              "xmlns:tns='http://ts.ch01/' " +
              "xmlns:xsd='http://www.w3.org/2001/XMLSchema'> " +
            "<soap:Body>" +
            "<tns:getTimeAsElapsed xsi:nil='true'/>" +
            "</soap:Body>" +
            "</soap:Envelope>";

        Map<String, Object> request_context = dispatch.getRequestContext();
        request_context.put(MessageContext.HTTP_REQUEST_METHOD, "POST");
        StreamSource source = make_stream_source(soap_request);
        Source result = dispatch.invoke(source);
        display_result(result, ns_URI.toString());
    }

    private void display_result(Source result, String uri) {
        DOMResult dom_result = new DOMResult();
        try {
            Transformer trans = TransformerFactory.newInstance().newTransformer();
            trans.transform(result, dom_result);

            XPathFactory xpf = XPathFactory.newInstance();
            XPath xp = xpf.newXPath();
            xp.setNamespaceContext(new NSResolver("tns", uri));
            // In original version, "//time_result" instead
            String result_string = xp.evaluate("//return", dom_result.getNode());
            System.out.println(result_string);
        }
        catch(TransformerConfigurationException e) { System.err.println(e); }
        catch(TransformerException e) { System.err.println(e); }
        catch(XPathExpressionException e) { System.err.println(e); }
    }

    private StreamSource make_stream_source(String msg) {
        ByteArrayInputStream stream = new ByteArrayInputStream(msg.getBytes());
        return new StreamSource(stream);
    }
}
```

The SOAP request document is hardcoded as a string. The rest of the setup is straight-forward. After a service object is created and a port added with the `TimeServer`'s end-point, a `Dispatch` proxy is created with a `Service.Mode.PAYLOAD` so that the SOAP request document becomes an XML `Source` transported to the service in the body of the HTTP request. The SOAP-based service responds with a SOAP envelope, which an `XPath` object then searches for the integer value that gives the elapsed milliseconds. On a sample run, the output was 1,214,514,573,623 (with commas added for readability) on a RESTful call to `getTimeAsElapsed`.

Implementing RESTful Web Services As HttpServlets

Here is a short review of servlets with emphasis on their use to deliver RESTful services. The class `HttpServlet` extends the class `GenericServlet`, which in turn implements the `Servlet` interface. All three are in the package `javax.servlet`, which is not included in core Java. The `Servlet` interface declares five methods, the most important of which is the `service` method that a web container invokes on every request to a servlet. The `service` method has a `ServletRequest` and a `ServletResponse` parameter. The request is a map that contains the request information from a client, and the response provides a network connection back to the client. The `GenericServlet` class implements the `Service` methods in a transport-neutral fashion, whereas its `HttpServlet` subclass im-plements these methods in an HTTP-specific way. Accordingly, the service parame-ters in the `HttpServlet` have the types `HttpServletRequest` and `HttpServletResponse`. The `HttpServlet` also provides request filtering: the service method dispatches an in-coming GET request to the method `doGet`, an incoming POST request to the method `doPost`, and so on. Figure 4-2 depicts a servlet container with several servlets.

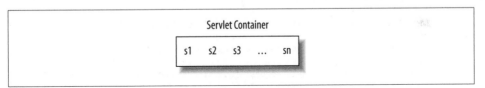

Figure 4-2. A servlet container with instances of various servlets

In the `HttpServlet` class, the `do` methods are no-ops (that is, methods with empty bod-ies) that can be overridden as needed in a programmer-derived subclass. For example, if the class `MyServlet` extends `HttpServlet` and overrides `doGet` but not `doPost`, then `doPost` remains a no-op in `MyServlet` instances.

`HttpServlet`s are a natural, convenient way to implement RESTful web services for two reasons. First, such servlets provide methods such as `doGet` and `doDelete` that match up with HTTP verbs, and these methods execute as callbacks that the web container invokes as needed. Second, the `HttpServletRequest` and `HttpServletResponse` are the same two arguments to every `do` method, which encourages a uniform pattern of request processing: client-supplied data are read from the `HttpServletRequest` map and processed as required; then a response is sent back to the client through the output stream associated with the `HttpServletResponse`.

The RabbitCounterServlet

The `RabbitCounterServlet` that follows is a RESTful, servlet-based version of the SOAP-based `RabbitCounter` service of Chapter 3. The service has a deliberately simple logic to keep focus on what makes the servlet such an attractive implementation of a RESTful service. Here is the source code:

```
package ch04.rc;

import javax.servlet.ServletException;
import javax.servlet.http.HttpServlet;
import javax.servlet.http.HttpServletRequest;
import javax.servlet.http.HttpServletResponse;
import javax.xml.ws.http.HTTPException;
import java.util.Collections;
import java.util.Map;
import java.util.HashMap;
import java.util.Collection;
import java.util.List;
import java.util.ArrayList;
import java.io.IOException;
import java.io.ByteArrayInputStream;
import java.io.ByteArrayOutputStream;
import java.io.OutputStream;
import java.beans.XMLEncoder;

public class RabbitCounterServlet extends HttpServlet {
    private Map<Integer, Integer> cache;

    // Executed when servlet is first loaded into container.
    public void init() {
        cache = Collections.synchronizedMap(new HashMap<Integer, Integer>());
    }
    public void doGet(HttpServletRequest request, HttpServletResponse response) {
        String num = request.getParameter("num");

        // If no query string, assume client wants the full list
        if (num == null) {
            Collection<Integer> fibs = cache.values();
            send_typed_response(request, response, fibs);
        }
        else {
            try {
                Integer key = Integer.parseInt(num.trim());
                Integer fib = cache.get(key);
                if (fib == null) fib = -1;
                send_typed_response(request, response, fib);
            }
```

```
            catch(NumberFormatException e) {
                send_typed_response(request, response, -1);
            }
        }
    }

    public void doPost(HttpServletRequest request, HttpServletResponse response) {
        String nums = request.getParameter("nums");
        if (nums == null)
            throw new HTTPException(HttpServletResponse.SC_BAD_REQUEST);

        // Extract the integers from a string such as: "[1, 2, 3]"
        nums = nums.replace('[', '\0');
        nums = nums.replace(']', '\0');
        String[ ] parts = nums.split(", ");
        List<Integer> list = new ArrayList<Integer>();
        for (String next : parts) {
            int n = Integer.parseInt(next.trim());
            cache.put(n, countRabbits(n));
            list.add(cache.get(n));
        }
        send_typed_response(request, response, list + " added.");
    }

    public void doDelete(HttpServletRequest request, HttpServletResponse response) {
        String key = request.getParameter("num");
        // Only one Fibonacci number may be deleted at a time.
        if (key == null)
            throw new HTTPException(HttpServletResponse.SC_BAD_REQUEST);
        try {
            int n = Integer.parseInt(key.trim());
            cache.remove(n);
            send_typed_response(request, response, n + " deleted.");
        }
        catch(NumberFormatException e) {
            throw new HTTPException(HttpServletResponse.SC_BAD_REQUEST);
        }
    }
    public void doPut(HttpServletRequest req, HttpServletResponse res) {
        throw new HTTPException(HttpServletResponse.SC_METHOD_NOT_ALLOWED);
    }

    public void doInfo(HttpServletRequest req, HttpServletResponse res) {
        throw new HTTPException(HttpServletResponse.SC_METHOD_NOT_ALLOWED);
    }

    public void doHead(HttpServletRequest req, HttpServletResponse res) {
        throw new HTTPException(HttpServletResponse.SC_METHOD_NOT_ALLOWED);
    }

    public void doOptions(HttpServletRequest req, HttpServletResponse res) {
        throw new HTTPException(HttpServletResponse.SC_METHOD_NOT_ALLOWED);
    }
```

```java
        private void send_typed_response(HttpServletRequest request,
                                         HttpServletResponse response,
                                         Object data) {
            String desired_type = request.getHeader("accept");

            // If client requests plain text or HTML, send it; else XML.
            if (desired_type.contains("text/plain"))
                send_plain(response, data);
            else if (desired_type.contains("text/html"))
                send_html(response, data);
            else
                send_xml(response, data);
        }

    // For simplicity, the data are stringified and then XML encoded.
    private void send_xml(HttpServletResponse response, Object data) {
        try {
            XMLEncoder enc = new XMLEncoder(response.getOutputStream());
            enc.writeObject(data.toString());
            enc.close();
        }
        catch(IOException e) {
            throw new HTTPException(HttpServletResponse.SC_INTERNAL_SERVER_ERROR);
        }
    }
    private void send_html(HttpServletResponse response, Object data) {
        String html_start =
            "<html><head><title>send_html response</title></head><body><div>";
        String html_end = "</div></body></html>";
        String html_doc = html_start + data.toString() + html_end;
        send_plain(response, html_doc);
    }

    private void send_plain(HttpServletResponse response, Object data) {
        try {
            OutputStream out = response.getOutputStream();
            out.write(data.toString().getBytes());
            out.flush();
        }
        catch(IOException e) {
            throw new HTTPException(HttpServletResponse.SC_INTERNAL_SERVER_ERROR);
        }
    }

    private int countRabbits(int n) {
        if (n < 0) throw new HTTPException(403);

        // Easy cases.
        if (n < 2) return n;
        // Return cached value if present.
        if (cache.containsKey(n)) return cache.get(n);
        if (cache.containsKey(n - 1) && cache.containsKey(n - 2)) {
            cache.put(n, cache.get(n - 1) + cache.get(n - 2));
            return cache.get(n);
        }
```

```
    // Otherwise, compute from scratch, cache, and return.
    int fib = 1, prev = 0;
    for (int i = 2; i <= n; i++) {
        int temp = fib;
        fib += prev;
        prev = temp;
    }
    cache.put(n, fib);
    return fib;
  }
}
```

Compiling and Deploying a Servlet

The `RabbitCounterServlet` cannot be compiled using only core Java. The container Apache Tomcat, typically shortened to Tomcat, is the reference implementation for servlet containers and is available for free download at *http://tomcat.apache.org*. The current version is 6.x. For simplicity, assume that *TOMCAT_HOME* is the install directory. Immediately under *TOMCAT_HOME* are three subdirectories of interest:

TOMCAT_HOME/bin

This subdirectory contains *startup* and *shutdown* scripts for Unix and Windows. To start Tomcat, execute the *startup* script. To test whether Tomcat started, open a browser to the URL *http://localhost:8080*.

TOMCAT_HOME/logs

Tomcat maintains various log files, which are helpful for determining whether a servlet deployed successfully.

TOMCAT_HOME/webapps

Servlets are deployed as *WAR* (Web ARchive) files, which are JAR files with a *.war* extension. To deploy a servlet, copy its WAR file to this directory. A detailed explanation follows.

The source code for the `RabbitCounterServlet` is in the package `ch04.rc`. Here are the steps for compiling, packaging, and deploying the servlet. The compilation command occurs at the working directory, which has *ch04* as a subdirectory:

- The servlet must be compiled with various packages that Tomcat supplies. Here is the command from the working directory:

    ```
    % javac -cp .:TOMCAT_HOME/lib/servlet-api.jar ch04/rc/*.java
    ```

    On Unix-like systems, it is `$TOMCAT_HOME` and on Windows systems, it is `%TOMCAT_HOME%`. Also, Unix-like systems use the colon to separate items on the classpath, whereas Windows uses the semicolon.

- In a deployed WAR file, Tomcat expects to find *.class* files such as the compiled `RabbitCounterServlet` in the tree rooted at *WEB-INF/classes*. In the working directory, create the subdirectory *WEB-INF/classes/ch04/rc* and then copy the compiled servlet into this subdirectory.

- Almost every production-grade servlet has a configuration file named *WEB-INF/web.xml*, although technically this is no longer a requirement. Here is the configuration file for this example:

```xml
<?xml version="1.0" encoding="UTF-8"?>
<web-app
    xmlns="http://java.sun.com/xml/ns/j2ee"
    xmlns:xsi="http://www.w3.org/2001/XMLSchema-instance"
    xsi:schemaLocation="http://java.sun.com/xml/ns/j2ee
        http://java.sun.com/xml/ns/j2ee/web-app_2_4.xsd"
        version="2.4">
    <display-name>RabbitCounter Servlet</display-name>
    <servlet>
        <servlet-name>rcounter</servlet-name>
        <servlet-class>
          ch04.rc.RabbitCounterServlet
        </servlet-class>
        <load-on-startup>0</load-on-startup>
    </servlet>
    <servlet-mapping>
        <servlet-name>rcounter</servlet-name>
        <url-pattern>/fib</url-pattern>
    </servlet-mapping>
</web-app>
```

    Of interest now is the `url-pattern` section, which gives `/fib` as the pattern. Assume that the servlet resides in the WAR file *rc.war*, as will be shown shortly. In this case, the URL for the servlet is *http://localhost:8080/rc/fib*. The configuration document recommends that Tomcat not load a servlet instance when the servlet is first deployed but rather to wait until the first client request.

- The servlet is now ready to be packaged and deployed. From the working directory, the command:

```
% jar cvf rc.war WEB-INF
```

    creates the WAR file, which holds the configuration file *web.xml* and the compiled servlet. Then deploy the WAR file by copying it to *TOMCAT_HOME/webapps*. A Tomcat log file can be inspected to see whether the deployment succeeded, or a browser can be opened to URL *http://localhost:8080/rc/fib*.

The `RabbitCounterServlet` overrides the `init` method, which the servlet container invokes when the servlet is first loaded. The method constructs the map that stores the Fibonacci numbers computed on a POST request. The other supported HTTP verbs are GET and DELETE. A GET request without a query string is treated as a request to read all of the numbers available, whereas a GET request with a query string is treated as a request for a specific Fibonacci number. The service allows the deletion of only one Fibonacci number at a time; hence, a DELETE request must have a query string that specifies which number to delete. The service does not implement the remaining CRUD operation, *update*; hence, the `doPut` method, like the remaining `do` methods, throws an HTTP 405 exception using the constant:

```
HttpServletResponse.SC_METHOD_NOT_ALLOWED
```

for clarity. There are similar constants for the other HTTP status codes.

Requests for MIME-Typed Responses

The RabbitCounterServlet differs from the first RESTful example in being implemented as a servlet instead of as a @WebServiceProvider. The RESTful servlet differs in a second way as well; that is, by honoring the request that a response be of a specified MIME type. Here is a client against the servlet:

```
import java.util.List;
import java.util.ArrayList;
import java.net.URL;
import java.net.HttpURLConnection;
import java.net.URLEncoder;
import java.net.MalformedURLException;
import java.net.URLEncoder;
import java.io.IOException;
import java.io.DataOutputStream;
import java.io.BufferedReader;
import java.io.InputStreamReader;

class ClientRC {
    private static final String url = "http://localhost:8080/rc/fib";

    public static void main(String[ ] args) {
        new ClientRC().send_requests();
    }

    private void send_requests() {
        try {
            HttpURLConnection conn = null;

            // POST request to create some Fibonacci numbers.
            List<Integer> nums = new ArrayList<Integer>();
            for (int i = 1; i < 15; i++) nums.add(i);
            String payload = URLEncoder.encode("nums", "UTF-8") + "=" +
                URLEncoder.encode(nums.toString(), "UTF-8");

            // Send the request
            conn = get_connection(url, "POST");
            conn.setRequestProperty("accept", "text/xml");
            DataOutputStream out = new DataOutputStream(conn.getOutputStream());
            out.writeBytes(payload);
            out.flush();
            get_response(conn);

            // GET to test whether POST worked
            conn = get_connection(url, "GET");
            conn.addRequestProperty("accept", "text/xml");
            conn.connect();
            get_response(conn);

            conn = get_connection(url + "?num=12", "GET");
            conn.addRequestProperty("accept", "text/plain");
```

```
                conn.connect();
                get_response(conn);

                // DELETE request
                conn = get_connection(url + "?num=12", "DELETE");
                conn.addRequestProperty("accept", "text/xml");
                conn.connect();
                get_response(conn);

                // GET request to test whether DELETE worked
                conn = get_connection(url + "?num=12", "GET");
                conn.addRequestProperty("accept", "text/html");
                conn.connect();
                get_response(conn);
            }
        catch(IOException e) { System.err.println(e); }
        catch(NullPointerException e) { System.err.println(e); }
    }

    private HttpURLConnection get_connection(String url_string, String verb) {
        HttpURLConnection conn = null;
        try {
            URL url = new URL(url_string);
            conn = (HttpURLConnection) url.openConnection();
            conn.setRequestMethod(verb);
            conn.setDoInput(true);
            conn.setDoOutput(true);
        }
        catch(MalformedURLException e) { System.err.println(e); }
        catch(IOException e) { System.err.println(e); }
        return conn;
    }
    private void get_response(HttpURLConnection conn) {
        try {
            String xml = "";
            BufferedReader reader =
                new BufferedReader(new InputStreamReader(conn.getInputStream()));
            String next = null;
            while ((next = reader.readLine()) != null)
                xml += next;
            System.out.println("The response:\n" + xml);
        }
        catch(IOException e) { System.err.println(e); }
    }
}
```

The client sends POST, GET, and DELETE requests, each of which specifies the desired MIME type of the response. For example, in the POST request the statement:

```
conn.setRequestProperty("accept", "text/xml");
```

requests that the response be an XML document. The value of the accept key need not be a single MIME type. For instance, the request statement could be:

```
conn.setRequestProperty("accept", "text/xml, text/xml, application/soap");
```

Indeed, the listed types can be prioritized and weighted for the service's consideration. This example sticks with single MIME types such as text/html.

The RabbitCounterServlet can send responses of MIME types text/xml (the default), text/html, and text/plain. Here is the client's output, slightly formatted and documented for readability:

```
The response: // from the initial POST request to create some Fibonacci numbers
<?xml version="1.0" encoding="UTF-8"?>
<java version="1.6.0_06" class="java.beans.XMLDecoder">
<string>
   [1, 1, 2, 3, 5, 8, 13, 21, 34, 55, 89, 144, 233, 377] added.
</string>
</java>

The response: // from a GET request to confirm that the POST worked
<?xml version="1.0" encoding="UTF-8"?>
<java version="1.6.0_06" class="java.beans.XMLDecoder">
<string>[1, 1, 2, 3, 5, 8, 13, 21, 34, 55, 89, 144, 233, 377]</string>
</java>

The response: // from a GET request with text/plain as the desired type
144

The response: // from a DELETE request
<?xml version="1.0" encoding="UTF-8"?>
<java version="1.6.0_06" class="java.beans.XMLDecoder">
<string>12 deleted.</string>
</java>

The response: // from a GET to confirm the DELETE with HTML as the desired type
<html><head><title>send_html response</title></head>
<body><div>-1</div></body>
</html>
```

In the last response, the returned value of −1 signals that the Fibonacci number for 12 is not available. The XML and HTML formats are simple, of course, but they illustrate how RESTful services can generate typed responses that satisfy requests.

Java Clients Against Real-World RESTful Services

There are many RESTful services available from well-known players such as Yahoo!, Amazon, and eBay, although controversy continues around the issue of what counts as a truly RESTful service. This section provides sample clients against some of these commercial REST-style services.

The Yahoo! News Service

Here HTTP verb is a client against Yahoo!'s RESTful service that summarizes the current news on a specified topic. The request is an HTTP GET with a query string:

```
import java.net.URI;
import java.util.Map;
import javax.xml.namespace.QName;
import javax.xml.ws.Service;
import javax.xml.ws.Dispatch;
import javax.xml.ws.http.HTTPBinding;
import javax.xml.transform.Source;
import javax.xml.transform.TransformerFactory;
import javax.xml.transform.Transformer;
import javax.xml.transform.dom.DOMResult;
import javax.xml.xpath.XPathFactory;
import javax.xml.xpath.XPath;
import javax.xml.xpath.XPathConstants;
import javax.xml.ws.handler.MessageContext;
import org.w3c.dom.NodeList;
import yahoo.NSResolver;

// A client against the Yahoo! RESTful news summary service.
class YahooClient {
    public static void main(String[ ] args) throws Exception {
        if (args.length < 1) {
            System.err.println("YahooClient <your AppID>");
            return;
        }
        String app_id = "appid=" + args[0];

        // Create a name for a service port.
        URI ns_URI = new URI("urn:yahoo:yn");
        QName serviceName = new QName("yahoo", ns_URI.toString());
        QName portName = new QName("yahoo_port", ns_URI.toString());

        // Now create a service proxy
        Service s = Service.create(serviceName);

        String qs = app_id + "&type=all&results=10&" +
                    "sort=date&language=en&query=quantum mechanics";

        // Endpoint address
        URI address = new URI("http",                       // HTTP scheme
                              null,                         // user info
                              "api.search.yahoo.com",       // host
                              80,                           // port
                              "/NewsSearchService/V1/newsSearch", // path
                              qs,                           // query string
                              null);                        // fragment

        // Add the appropriate port
        s.addPort(portName, HTTPBinding.HTTP_BINDING, address.toString());

        // From the service, generate a Dispatcher
        Dispatch<Source> d =
            s.createDispatch(portName, Source.class, Service.Mode.PAYLOAD);
        Map<String, Object> request_context = d.getRequestContext();
        request_context.put(MessageContext.HTTP_REQUEST_METHOD, "GET");
```

```
    // Invoke
    Source result = d.invoke(null);
    DOMResult dom_result = new DOMResult();
    Transformer trans = TransformerFactory.newInstance().newTransformer();
    trans.transform(result, dom_result);

    XPathFactory xpf = XPathFactory.newInstance();
    XPath xp = xpf.newXPath();
    xp.setNamespaceContext(new NSResolver("yn", ns_URI.toString()));
    NodeList resultList = (NodeList)
        xp.evaluate("/yn:ResultSet/yn:Result",
                    dom_result.getNode(),
                    XPathConstants.NODESET);

    int len = resultList.getLength();
    for (int i = 1; i <= len; i++) {
        String title =
            xp.evaluate("/yn:ResultSet/yn:Result(" + i + ")/yn:Title",
                        dom_result.getNode());
        String click =
            xp.evaluate("/yn:ResultSet/yn:Result(" + i + ")/yn:ClickUrl",
                        dom_result.getNode());
        System.out.printf("(%d) %s (%s)\n", i, title, click);
    }
  }
}
```

This client application expects, as a command-line argument, the user's application identifier. The news service is free but requires this identifier. (Signup is available at *http://www.yahoo.com*.) In this example, the client requests a maximum of 10 results on the topic of *quantum gravity*. Here is a segment of the raw XML that the Yahoo! service returns:

```
<?xml version="1.0" encoding="UTF-8"?>
<ResultSet xmlns:xsi="http://www.w3.org/2001/XMLSchema-instance"
    xmlns="urn:yahoo:yn"
    xsi:schemaLocation="urn:yahoo:yn
      http://api.search.yahoo.com/NewsSearchService/V1/NewsSearchResponse.xsd"
    totalResultsAvailable="9" totalResultsReturned="9"
    firstResultPosition="1">
<Result>
  <Title>Cosmic Log: Private space age turns 4</Title>
  <Summary>Science editor Alan Boyle's Weblog: Four years after the first
          private-sector spaceship crossed the 62-mile mark, some space-age
          dreams have been slow to rise while others have paid off.
  </Summary>
  <Url>
    http://cosmiclog.msnbc.msn.com/archive/2008/06/20/1158681.aspx
  </Url>
  <ClickUrl>
    http://cosmiclog.msnbc.msn.com/archive/2008/06/20/1158681.aspx
  </ClickUrl>
  <NewsSource>MSNBC</NewsSource>
  <NewsSourceUrl>http://www.msnbc.msn.com/</NewsSourceUrl>
  <Language>en</Language>
```

```
    <PublishDate>1213998337</PublishDate>
    <ModificationDate>1213998338</ModificationDate>
  </Result>
  ...
  </ResultSet>
```

Here is the parsed output that the YahooClient produces:

```
(1) Cosmic Log: Private space age turns 4 (http://cosmiclog.msnbc.msn.com/...
(2) Neutrino experiment shortcuts from novel to real world...
(3) There Will Be No Armageddon (http://www.spacedaily.com/reports/...
(4) TSX Venture Exchange Closing Summary for June 19, 2008 (http://biz.yahoo.com...
(5) Silver Shorts Reported (http://news.goldseek.com/GoldSeeker/1213848000.php)
(6) There will be no Armageddon (http://en.rian.ru/analysis/20080618/...
(7) New Lunar Prototype Vehicles Tested (Gallery)...
(8) World's Largest Quantum Bell Test Spans Three Swiss Towns...
(9) Creating science (http://www.michigandaily.com/news/2008/06/16/...
```

The client uses a Dispatch object to issue the request and an XPath object to search the DOM result for selected elements. The output above includes the Summary and the ClickURL elements from the raw XML. As *quantum gravity* is not a hot news topic, there were only 9 results from a request for 10.

The Yahoo! example underscores that clients of RESTful services assume the burden of processing the response document, which is typically XML, in some way that is appropriate to the application. Although there generally is an XML Schema that specifies how the raw XML is formatted, there is no service contract comparable to the WSDL used in SOAP-based services.

The Amazon E-Commerce Service: REST Style

Yahoo! exposes only RESTful web services, but Amazon provides its web services in two ways, as SOAP-based and as REST-style. The AmazonClientREST application that follows issues a *read* request against the Amazon E-Commerce service for books about the Fibonacci numbers. The client uses a Dispatch object and an HTTP GET request with a query string that specifies the details of the request:

```java
import java.util.Map;
import javax.xml.namespace.QName;
import javax.xml.ws.Service;
import javax.xml.ws.Dispatch;
import javax.xml.ws.http.HTTPBinding;
import javax.xml.transform.stream.StreamSource;
import javax.xml.transform.Source;
import javax.xml.transform.TransformerFactory;
import javax.xml.transform.Transformer;
import javax.xml.transform.dom.DOMResult;
import javax.xml.transform.TransformerConfigurationException;
import javax.xml.transform.TransformerException;
import javax.xml.xpath.XPathFactory;
import javax.xml.xpath.XPath;
import javax.xml.xpath.XPathConstants;
import javax.xml.xpath.XPathExpressionException;
```

```java
import javax.xml.ws.handler.MessageContext;
import org.w3c.dom.NodeList;
import org.w3c.dom.Node;
import ch04.dispatch.NSResolver;

class AmazonClientREST {
    private final static String uri =
        "http://webservices.amazon.com/AWSECommerceService/2005-03-23";

    public static void main(String[ ] args) throws Exception {
        if (args.length < 1) {
            System.err.println("Usage: AmazonClientREST <access key>");
            return;
        }
        new AmazonClientREST().item_search(args[0].trim());
    }

    private void item_search(String access_key) {
        QName service_name = new QName("awsREST", uri);
        QName port = new QName("awsPort", uri);

        String base_url = "http://ecs.amazonaws.com/onca/xml";
        String qs = "?Service=AWSECommerceService&" +
            "Version=2005-03-23&" +
            "Operation=ItemSearch&" +
            "ContentType=text%2Fxml&" +
            "AWSAccessKeyId=" + access_key + "&" +
            "SearchIndex=Books&" +
            "Keywords=Fibonacci";
        String endpoint = base_url + qs;

        // Now create a service proxy dispatcher.
        Service service = Service.create(service_name);
        service.addPort(port, HTTPBinding.HTTP_BINDING, endpoint);
        Dispatch<Source> dispatch =
            service.createDispatch(port, Source.class, Service.Mode.PAYLOAD);

        // Set HTTP verb.
        Map<String, Object> request_context = dispatch.getRequestContext();
        request_context.put(MessageContext.HTTP_REQUEST_METHOD, "GET");

        Source result = dispatch.invoke(null); // null payload: GET request
        display_result(result);
    }

    private void display_result(Source result) {
        DOMResult dom_result = new DOMResult();
        try {
            Transformer trans = TransformerFactory.newInstance().newTransformer();
            trans.transform(result, dom_result);
            XPathFactory xpf = XPathFactory.newInstance();
            XPath xp = xpf.newXPath();
            xp.setNamespaceContext(new NSResolver("aws", uri));
```

```
            NodeList authors = (NodeList)
                xp.evaluate("//aws:ItemAttributes/aws:Author",
                            dom_result.getNode(),
                            XPathConstants.NODESET);

            NodeList titles = (NodeList)
                xp.evaluate("//aws:ItemAttributes/aws:Title",
                            dom_result.getNode(),
                            XPathConstants.NODESET);

            int len = authors.getLength();
            for (int i = 0; i < len; i++) {
                Node author = authors.item(i);
                Node title  = titles.item(i);
                if (author != null && title != null) {
                    String a_name = author.getFirstChild().getNodeValue();
                    String t_name = title.getFirstChild().getNodeValue();
                    System.out.printf("%s: %s\n", a_name, t_name);
                }
            }
        }
        catch(TransformerConfigurationException e) { System.err.println(e); }
        catch(TransformerException e) { System.err.println(e); }
        catch(XPathExpressionException e) { System.err.println(e); }
    }
}
```

The response document is now raw XML rather than a SOAP envelope. However, the
raw XML conforms to the very same XML Schema document that is used in the
E-Commerce WSDL contract for the SOAP-based version of the service. In effect, then,
the only difference is that the raw XML is wrapped in a SOAP envelope in the case of
the SOAP-based service, but is simply the payload of the HTTP response in the case of
the REST-style service. Here is a segment of the raw XML:

```
<?xml version="1.0" encoding="UTF-8"?>
<ItemSearchResponse
    xmlns="http://webservices.amazon.com/AWSECommerceService/2005-03-23">
  ...
  <ItemSearchRequest>
    <Keywords>Fibonacci</Keywords>
    <SearchIndex>Books</SearchIndex>
  </ItemSearchRequest>
  ...
  <TotalResults>177</TotalResults>
  <TotalPages>18</TotalPages>
  ...
  <Items>
    <Item>
      <ItemAttributes>
        <Author>Carolyn Boroden</Author>
        <Manufacturer>McGraw-Hill</Manufacturer>
        <ProductGroup>Book</ProductGroup>
        <Title>Fibonacci Trading: How to Master Time and Price Advantage</Title>
      </ItemAttributes>
```

```
    </Item>
    ...
    </Items>
</ItemSearchResponse>
```

For variety, the `AmazonClientREST` parses the raw XML in a slightly different way than in earlier examples. In particular, the client uses XPath to get separate lists of authors and book titles:

```
NodeList authors = (NodeList) xp.evaluate("//aws:ItemAttributes/aws:Author",
                           dom_result.getNode(), XPathConstants.NODESET);
NodeList titles  = (NodeList)  xp.evaluate("//aws:ItemAttributes/aws:Title",
                           dom_result.getNode(), XPathConstants.NODESET);
```

and then loops through the lists to extract the author and the title using the DOM API. Here is the loop:

```
int len = authors.getLength();
for (int i = 0; i < len; i++) {
   Node author = authors.item(i);
   Node title  = titles.item(i);
   if (author != null && title != null) {
     String a_name = author.getFirstChild().getNodeValue();
     String t_name = title.getFirstChild().getNodeValue();
     System.out.printf("%s: %s\n", a_name, t_name);
   }
}
```

that produced, on a sample run, the following output:

```
Carolyn Boroden: Fibonacci Trading: How to Master Time and Price Advantage
Kimberly Elam: Geometry of Design: Studies in Proportion and Composition
Alfred S. Posamentier: The Fabulous Fibonacci Numbers
Ingmar Lehmann: Math for Mystics: From the Fibonacci sequence to Luna's Labyrinth...
Renna Shesso: Breakthrough Strategies for Predicting any Market...
Jeff Greenblatt: Wild Fibonacci: Nature's Secret Code Revealed
Joy N. Hulme: Fibonacci Analysis (Bloomberg Market Essentials: Technical Analysis)
Constance Brown: Fibonacci Fun: Fascinating Activities With Intriguing Numbers
Trudi Hammel Garland: Fibonacci Applications and Strategies for Traders
Robert Fischer: New Frontiers in Fibonacci Trading: Charting Techniques,...
```

The Amazon Simple Storage Service, known as Amazon S3, is a pay-for service also accessible through a SOAP-based or a RESTful client. As the name indicates, the service allows users to store and retrieve individual data objects, each of up to 5G in size. S3 is often cited as a fine example of a useful web service with a very simple interface.

The RESTful Tumblr Service

Perhaps the Tumblr service is best known for the associated term *tumblelog* or *tlog*, a variation on the traditional blog that emphasizes short text entries with associated multimedia such as photos, music, and film. It is common for tumblelogs to be artistic endeavors. The service is free but the full set of RESTful operations requires a user account, which can be set up at *http://www.tumblr.com*.

The `TumblrClient` that follows uses an `HttpURLConnection` to send a GET and a POST request against the Tumblr RESTful service. In this case, the `HttpURLConnection` is a better choice than `Dispatch` because a POST request to Tumblr does not contain an XML document but rather the standard key/value pairs. Here is the client code:

```java
import java.net.URL;
import java.net.HttpURLConnection;
import java.net.URLEncoder;
import java.net.MalformedURLException;
import java.net.URLEncoder;
import java.io.IOException;
import java.io.DataOutputStream;
import java.io.BufferedReader;
import java.io.InputStreamReader;
import javax.xml.transform.stream.StreamSource;
import javax.xml.transform.TransformerFactory;
import javax.xml.transform.Transformer;
import javax.xml.transform.TransformerConfigurationException;
import javax.xml.transform.TransformerException;
import javax.xml.transform.dom.DOMResult;
import javax.xml.xpath.XPathFactory;
import javax.xml.xpath.XPath;
import javax.xml.xpath.XPathConstants;
import javax.xml.xpath.XPathExpressionException;
import java.io.ByteArrayInputStream;
import org.w3c.dom.NodeList;
import org.w3c.dom.Node;

class TumblrClient {
    public static void main(String[ ] args) {
        if (args.length < 2) {
            System.err.println("Usage: TumblrClient <email> <passwd>");
            return;
        }
        new TumblrClient().tumble(args[0], args[1]);
    }

    private void tumble(String email, String password) {
        try {
            HttpURLConnection conn = null;

            // GET request.
            String url = "http://mgk-cdm.tumblr.com/api/read";
            conn = get_connection(url, "GET");
            conn.setRequestProperty("accept", "text/xml");
            conn.connect();
            String xml = get_response(conn);
            if (xml.length() > 0) {
                System.out.println("Raw XML:\n" + xml);
                parse(xml, "\nSki photo captions:", "//photo-caption");
            }

            // POST request
            url = "http://www.tumblr.com/api/write";
            conn = get_connection(url, "POST");
```

```java
            String title = "Summer thoughts up north";
            String body = "Craigieburn Ski Area, NZ";
            String payload =
                URLEncoder.encode("email", "UTF-8") + "=" +
                URLEncoder.encode(email, "UTF-8") + "&" +
                URLEncoder.encode("password", "UTF-8") + "=" +
                URLEncoder.encode(password, "UTF-8") + "&" +
                URLEncoder.encode("type", "UTF-8") + "=" +
                URLEncoder.encode("regular", "UTF-8") + "&" +
                URLEncoder.encode("title", "UTF-8") + "=" +
                URLEncoder.encode(title, "UTF-8") + "&" +
                URLEncoder.encode("body", "UTF-8") + "=" +
                URLEncoder.encode(body, "UTF-8");
            DataOutputStream out = new DataOutputStream(conn.getOutputStream());
            out.writeBytes(payload);
            out.flush();
            String response = get_response(conn);
            System.out.println("Confirmation code: " + response);
        }
        catch(IOException e) { System.err.println(e); }
        catch(NullPointerException e) { System.err.println(e); }
    }

    private HttpURLConnection get_connection(String url_s, String verb) {
        HttpURLConnection conn = null;
        try {
            URL url = new URL(url_s);
            conn = (HttpURLConnection) url.openConnection();
            conn.setRequestMethod(verb);
            conn.setDoInput(true);
            conn.setDoOutput(true);
        }
        catch(MalformedURLException e) { System.err.println(e); }
        catch(IOException e) { System.err.println(e); }
        return conn;
    }

    private String get_response(HttpURLConnection conn) {
        String xml = "";
        try {
            BufferedReader reader =
                new BufferedReader(new InputStreamReader(conn.getInputStream()));
            String next = null;
            while ((next = reader.readLine()) != null) xml += next;
        }
        catch(IOException e) { System.err.println(e); }
        return xml;
    }

    private void parse(String xml, String msg, String pattern) {
        StreamSource source =
            new StreamSource(new ByteArrayInputStream(xml.getBytes()));
        DOMResult dom_result = new DOMResult();
        System.out.println(msg);
```

```
        try {
            Transformer trans = TransformerFactory.newInstance().newTransformer();
            trans.transform(source, dom_result);
            XPathFactory xpf = XPathFactory.newInstance();
            XPath xp = xpf.newXPath();
            NodeList list = (NodeList)
                xp.evaluate(pattern, dom_result.getNode(), XPathConstants.NODESET);
            int len = list.getLength();
            for (int i = 0; i < len; i++) {
                Node node = list.item(i);
                if (node != null)
                  System.out.println(node.getFirstChild().getNodeValue());
            }
        }
        catch(TransformerConfigurationException e) { System.err.println(e); }
        catch(TransformerException e) { System.err.println(e); }
        catch(XPathExpressionException e) { System.err.println(e); }
    }
}
```

The URL for the GET request is:

```
http://mgk-cdm.tumblr.com/api/read
```

which is the URL for my Tumblr account's site with **api/read** appended. The request returns all of my public (that is, unsecured) postings. Here is part of the raw XML returned as the response:

```
<?xml version="1.0" encoding="UTF-8"?>
<tumblr version="1.0">
  <tumblelog name="mgk-cdm" timezone="US/Eastern" title="Untitled"/>
  <posts start="0" total="5">
    <post id="40130991" url="http://mgk-cdm.tumblr.com/post/40130991" type="photo"
        date-gmt="2008-06-28 03:09:29 GMT" date="Fri, 27 Jun 2008 23:09:29"
        unix-timestamp="1214622569">
      <photo-caption>Trying the new skis, working better than I am.</photo-caption>
      <photo-url max-width="500">
        http://media.tumblr.com/awK1GiaTRar6p46p6Xy13mBH_500.jpg
      </photo-url>
    </post>
    ...
    <post id="40006745" url="http://mgk-cdm.tumblr.com/post/40006745" type="regular"
        date-gmt="2008-06-27 04:12:53 GMT" date="Fri, 27 Jun 2008 00:12:53"
        unix-timestamp="1214539973">
      <regular-title>Weathering the weather</regular-title>
      <regular-body>miserable, need to get fully wet or not at all</regular-body>
    </post>
    ...
    <post id="40006638" url="http://mgk-cdm.tumblr.com/post/40006638" type="regular"
        date-gmt="2008-06-27 04:11:34 GMT" date="Fri, 27 Jun 2008 00:11:34"
        unix-timestamp="1214539894">
      <regular-title>tumblr. API</regular-title>
      <regular-body>Very restful</regular-body>
    </post>
```

```
    </posts>
  </tumblr>
```

The raw XML has a very simple structure, dispensing even with namespaces. The `TumblrClient` uses XPath to extract a list of the photo captions:

```
Ski photo captions:
Trying the new skis, working better than I am.
Very tough day on the trails; deep snow, too deep for skating.
Long haul up, fun going down.
```

The client then sends a POST request, which adds a new entry in my Tumblr posts. The URL now changes to the main Tumblr site, *http://www.tumblr.com*, with */api/ write* appended. My email and password must be included in the POST request's payload, which the following code segment handles:

```
String payload =
    URLEncoder.encode("email", "UTF-8") + "=" +
    URLEncoder.encode(email, "UTF-8") + "&" +
    URLEncoder.encode("password", "UTF-8") + "=" +
    URLEncoder.encode(password, "UTF-8") + "&" +
    URLEncoder.encode("type", "UTF-8") + "=" +
    URLEncoder.encode("regular", "UTF-8") + "&" +
    URLEncoder.encode("title", "UTF-8") + "=" +
    URLEncoder.encode(title, "UTF-8") + "&" +
    URLEncoder.encode("body", "UTF-8") + "=" +
    URLEncoder.encode(body, "UTF-8");
DataOutputStream out = new DataOutputStream(conn.getOutputStream());
out.writeBytes(payload);
out.flush();
```

The documentation for the Tumblr API is a single page. The API supports the CRUD *read* and *create* operations through the */api/read* and the */api/write* suffixes. As usual, a *read* operation is done through a GET request, and a *create* operation is done through a POST request. Tumblr does support some variation. For example, the suffix */api/ read/json* causes the response to be *JSON* (JavaScript Object Notation) instead of XML. An HTTP POST to the Tumblr site can be used to upload images, audio, and film in addition to text postings, and multimedia may be uploaded as either unencoded bytes or as standard URL-encoded payloads in the POST request's body.

The simplicity of the Tumblr API encourages the building of graphical interfaces and plugins that, in turn, allow Tumblr to interact easily with other sites such as Facebook. The Tumblr API is a fine example of how much can be done with so little.

WADLing with Java-Based RESTful Services

In SOAP-based web services, the WSDL document is a blessing to programmers because this service contract can be used to generate client-side artifacts and, indeed, even a service-side interface. RESTful services do not have an official or even widely accepted counterpart to the WSDL, although there are efforts in that direction. Among them is the WADL initiative (*https://wadl.dev.java.net*). WADL stands for Web Application Description Language.

The WADL download includes the *wadl2java* utility, a library of required JAR files, and a sample WADL file named *YahooSearch.wadl*. The download also has Ant, Maven, and command-line scripts for convenience. To begin, here is a Yahoo! client that uses *wadl2java*-generated artifacts:

```
import com.yahoo.search.ResultSet;
import com.yahoo.search.ObjectFactory;
import com.yahoo.search.Endpoint;
import com.yahoo.search.Endpoint.NewsSearch;
import com.yahoo.search.Type;
import com.yahoo.search.Result;
import com.yahoo.search.Sort;
import com.yahoo.search.ImageType;
import com.yahoo.search.Output;
import com.yahoo.search.Error;
import com.yahoo.search.SearchErrorException;
import javax.xml.bind.JAXBException;
import java.io.IOException;
import java.util.List;

class YahooWADL {
    public static void main(String[ ] args) {
        if (args.length < 1) {
            System.err.println("Usage: YahooWADL <app id>");
            return;
        }
        String app_id = args[0];
        try {
            NewsSearch service = new NewsSearch();
            String query = "neutrino";

            ResultSet result_set = service.getAsResultSet(app_id, query);
            List<Result> list = result_set.getResultList();
            int i = 1;
            for (Result next : list) {
                String title = next.getTitle();
                String click = next.getClickUrl();
                System.out.printf("(%d) %s %s\n", i++, title, click);
            }
        }
        catch(JAXBException e) { System.err.println(e); }
        catch(SearchErrorException e) { System.err.println(e); }
        catch(IOException e) { System.err.println(e); }
    }
}
```

The code is cleaner than my original `YahooClient`. The *wadl2java*-generated code hides the XML processing and other grimy details such as the formatting of an appropriate query string for a GET request against the Yahoo! News Service. The client-side artifacts also include utility classes for getting images from the Yahoo! service. On a sample run, the `YahooWADL` client produced this output on a request for articles that include the keyword *neutrino*:

```
(1) Congress to the rescue for Fermi jobs http://www.dailyherald.com/story/...
(2) AIP FYI #69: Senate FY 2009 National Science Foundation Funding Bill...
(3) Linked by Thom Holwerda on Wed 12th Sep 2007 11:51 UTC...
(4) The World's Nine Largest Science Projects http://science.slashdot.org/...
(5) Funding bill may block Fermi layoffs http://www.suntimes.com/business/...
(6) In print http://www.sciencenews.org/view/generic/id/33654/title/For_Kids...
(7) Recent Original Stories http://www.osnews.com/thread?284017
(8) Antares : un télescope pointé vers le sol qui utilise la terre comme filtre...
(9) Software addresses quality of hands-free car phone audio...
(10) Planetary science: Tunguska at 100 http://www.nature.com/news/2008/...
```

Here is the WADL document used to generate the client-side artifacts:

```xml
<?xml version="1.0"?>
<!--
The contents of this file are subject to the terms of the Common Development and
Distribution License (the "License").  You may not use this file except in
compliance with the License. You can obtain a copy of the license at
     http://www.opensource.org/licenses/cddl1.php
See the License for the specific language governing
permissions and limitations under the License.
-->
<application xmlns:xsd="http://www.w3.org/2001/XMLSchema"
             xmlns:yn="urn:yahoo:yn"
             xmlns:ya="urn:yahoo:api"
             xmlns:html="http://www.w3.org/1999/xhtml"
             xmlns="http://research.sun.com/wadl/2006/10">
  <grammars>
    <include href="NewsSearchResponse.xsd"/>
    <include href="NewsSearchError.xsd"/>
  </grammars>

  <resources base="http://api.search.yahoo.com/NewsSearchService/V1/">
    <resource path="newsSearch">
      <doc xml:lang="en" title="Yahoo News Search Service">
        The <html:i>Yahoo News Search</html:i> service provides online
                    searching of news stories from around the world.
      </doc>
      <param name="appid" type="xsd:string" required="true" style="query">
        <doc>The application ID. See
        <html:a href="http://developer.yahoo.com/faq/index.html#appid">
          Application IDs
        </html:a> for more information.
        </doc>
      </param>
      <method href="#search"/>
    </resource>
  </resources>

  <method name="GET" id="search">
    <doc xml:lang="en" title="Search news stories by keyword"/>
    <request>
      <param name="query" type="xsd:string" required="true" style="query">
```

```
      <doc xml:lang="en" title="Space separated keywords to search for"/>
    </param>
    <param name="type" type="xsd:string" default="all" style="query">
      <doc xml:lang="en" title="Keyword matching"/>
      <option value="all">
        <doc>All query terms.</doc>
      </option>
      <option value="any">
        <doc>Any query terms.</doc>
      </option>
      <option value="phrase">
        <doc>Query terms as a phrase.</doc>
      </option>
    </param>
    <param name="results" type="xsd:int" default="10" style="query">
      <doc xml:lang="en" title="Number of results"/>
    </param>
    <param name="start" type="xsd:int" default="1" style="query">
      <doc xml:lang="en" title="Index of first result"/>
    </param>
    <param name="sort" type="xsd:string" default="rank" style="query">
      <doc xml:lang="en" title="Sort by date or rank"/>
      <option value="rank"/>
      <option value="date"/>
    </param>
    <param name="language" type="xsd:string" style="query">
      <doc xml:lang="en" title="Language filter, omit for any language"/>
    </param>
    <param name="output" type="xsd:string" default="xml" style="query">
      <doc>The format for the output. If <html:em>json</html:em> is requested,
          the results will be returned in
          <html:a href="http://developer.yahoo.com/common/json.html">
              JSON
          </html:a> format. If <html:em>php</html:em> is requested, the
          results will be returned in
           <html:a href="http://developer.yahoo.com/common/phpserial.html">
              Serialized PHP
           </html:a> format.
      </doc>
      <option value="xml"/>
      <option value="json"/>
      <option value="php"/>
    </param>
    <param name="callback" type="xsd:string" style="query">
      <doc>The name of the callback function to wrap around the JSON data.
          The following characters are allowed: A-Z a-z 0-9 .
          [ ] and _. If output=json has not been requested, this
          parameter has no effect. More information on the callback can be
          found in the
          <html:a href="http://developer.yahoo.com/common/json.html
          #callbackparam">Yahoo! Developer Network JSON
          Documentation</html:a>.
      </doc>
    </param>
  </request>
```

```
        <response>
          <representation mediaType="application/xml" element="yn:ResultSet">
            <doc xml:lang="en" title="A list of news items matching the query"/>
          </representation>
          <fault id="SearchError" status="400" mediaType="application/xml"
                    element="ya:Error"/>
        </response>
      </method>
    </application>
```

The WADL document begins with references to two XSD documents, one of which is the grammar for error documents and the other of which is the grammar for normal response documents from the service. Next comes a list of available resources, in this case only the news search service. The `methods` section describes the HTTP verbs and, by implication, the CRUD operations that can be used against the service. In the case of the Yahoo! news service, only GET requests are allowed. The remaining sections provide details about the parameters that can accompany requests and responses. XSD type information of the sort found in WSDL documents occurs throughout the WADL as well.

Executing the *wadl2java* utility on the *YahooSearch.wadl* file generates 11 Java source files, the most important of which for the client-side programmer is *Endpoint.java*. The `Endpoint` class encapsulates the `static` class `NewsSearch`, an instance of which has utility methods such as `getAsResultSet`. For reference, here is the main segment of the client `YahooWADL` shown earlier. The WADL artifacts allow the code to be short and clear:

```
NewsSearch service = new NewsSearch();
String query = "neutrino";

ResultSet result_set = service.getAsResultSet(app_id, query);
List<Result> list = result_set.getResultList();
int i = 1;
for (Result next : list) {
    String title = next.getTitle();
    String click = next.getClickUrl();
    System.out.printf("(%d) %s %s\n", i++, title, click);
}
```

The search string, given in this case with the object reference `query`, is just a list of keywords separated by blanks. The `NewsSearch` object has properties for specifying sort order, the maximum number of items to return, and the like. The generated artifacts do lighten the coding burden.

WADL has stirred interest and discussion but remains, for now, a Java-centric initiative. The critical issue is whether REST-style services can be standardized to the extent that utilities such as *wadl2java* and perhaps *java2wadl* can measure up to the utilities now available for SOAP-based services. If it is fair to criticize SOAP-based services for being over-engineered, it is also fair to fault REST-style services for being under-engineered.

Problems in the wadl2java-Generated Code

Among the files that the *wadl2java* utility generates are *Error.java* and *ObjectFactory.java*. In each file, any occurrences of `urn:yahoo:api` should be changed to `urn:yahoo:yn`. In the 1.0 distribution, there were two occurrences in each source file. Without these changes, a JAX-B exception is thrown when the results from the Yahoo! search are un-marshaled into Java objects. The changes could be made to the XSD document and the WADL document that the *wadl2java* utility uses.

JAX-RS: WADLing Through Jersey

Jersey is the centerpiece project for the recent *JAX-RS* (Java API for XML-RESTful Web Services). Jersey applications can be deployed through familiar commercial-grade containers such as standalone Tomcat and GlassFish, but Jersey also provides the lightweight Grizzly container that is well suited for learning the framework. Jersey works well with Maven. A deployed Jersey service automatically generates a WADL, which is then available through a standard GET request. A good place to start is *https://jersey.dev.java.net*.

A Jersey service adheres to REST principles. A service accepts the usual RESTful requests for CRUD operations specified with the standard HTTP verbs GET, POST, DELETE, and PUT. A request is targeted at a Jersey resource, which is a POJO. Here is the `MsgResource` class to illustrate:

```
package msg.resources;

import javax.ws.rs.Path;
import javax.ws.rs.PathParam;
import javax.ws.rs.FormParam;
import javax.ws.rs.Produces;
import javax.ws.rs.GET;
import javax.ws.rs.POST;
import javax.ws.rs.DELETE;
import java.beans.XMLEncoder;
import java.io.ByteArrayOutputStream;

// This is the base path, which can be extended at the method level.
@Path("/")
public class MsgResource {
    private static String msg = "Hello, world!";

    @GET
    @Produces("text/plain")
    public String read() {
        return msg + "\n";
    }

    @GET
    @Produces("text/plain")
    @Path("{extra}")
```

```
    public String personalized_read(@PathParam("extra") String cus) {
        return this.msg + ": " + cus + "\n";
    }

    @POST
    @Produces("text/xml")
    public String create(@FormParam("msg") String new_msg ) {
        this.msg = new_msg;

        ByteArrayOutputStream stream = new ByteArrayOutputStream();
        XMLEncoder enc = new XMLEncoder(stream);
        enc.writeObject(new_msg);
        enc.close();
        return new String(stream.toByteArray()) + "\n";
    }

    @DELETE
    @Produces("text/plain")
    public String delete() {
        this.msg = null;
        return "Message deleted.\n";
    }
}
```

The class has intuitive annotations, including the ones for the HTTP verbs and the response MIME types. The @Path annotation right above the MsgResource class declaration is used to decouple the resource from any particular base URL. For example, the MsgResource might be available at the base URL *http://foo.bar.org:1234*, at the base URL *http://localhost:9876*, and so on. The @GET, @POST, and @DELETE annotations specify the appropriate HTTP verb for a particular service operation. The @Produces annotation specifies the MIME type of the response, in this case either text/plain for the GET and DELETE operations or text/xml for the POST operation. Each method annotated as a MsgResource is responsible for generating the declared response type.

The MsgResource class could be put in a WAR file along with the supporting Jersey JAR files and then deployed in a servlet container such as Tomcat. There is, however, a quick way to publish a Jersey service during development. Here is the publisher class to illustrate:

```
import com.sun.jersey.api.container.grizzly.GrizzlyWebContainerFactory;
import java.util.Map;
import java.util.HashMap;

class JerseyPublisher {
    public static void main(String[ ] args) {
        final String base_url = "http://localhost:9876/";
        final Map<String, String> config = new HashMap<String, String>();

        config.put("com.sun.jersey.config.property.packages",
                   "msg.resources"); // package with resource classes

        System.out.println("Grizzly starting on port 9876.\n" +
                           "Kill with Control-C.\n");
```

```
            try {
                GrizzlyWebContainerFactory.create(base_url, config);
            }
            catch(Exception e) { System.err.println(e); }
        }
    }
```

Grizzly requires configuration information about the package, in this case named `msg.resources`, that contains the resources available to clients. In this example, the package houses only the single class `MsgResource` but could house several resources. On each incoming request, the Grizzly container surveys the available resources to determine which method should handle the request. RESTful routing is thus in effect. For example, a POSTed request is delegated only to a method annotated with `@POST`.

Compiling and executing the resource and the publisher requires that several Jersey JAR files be on the classpath. Here is the list of five under the current release:

```
asm-3.1.jar
grizzly-servlet-webserver-1.8.3.jar
jersey-core-0.9-ea.jar
jersey-server-0.9-ea.jar
jsr311-api-0.9.jar
```

All of these JAR files, together with others for a Maven-centric version of Jersey, are available at the Jersey home page cited earlier.

Once the `JerseyPublisher` has been started, a browser or a utility such as *curl* can be used to access the resource. For example, the *curl* command:

```
% curl http://localhost:9876/
```

issues a GET request against the service, which causes the `@GET`-annotated `read` method to execute. The response is the default message:

```
Hello, world!
```

By contrast, the *curl* command:

```
% curl -d msg='Goodbye, cruel world!' http://localhost:9876/echo/fred
```

issues a POST request against the service, which in turn causes the `@POST`-annotated `create` method to execute. (A REST purist might argue that a PUT operation would be more appropriate here, as the `create` method arguably updates an existing message rather than creates a message.) The `create` method uses the `@FormParam` annotation so that the POSTed data are available as the method's argument. The `@FormParam` parameter, in this case the string `msg`, need not be the same as the method parameter, in this case `new_msg`. The output is:

```
<?xml version="1.0" encoding="UTF-8"?>
<java version="1.6.0_06" class="java.beans.XMLDecoder">
 <string>Goodbye, cruel world!</string>
</java>
```

because the @Produces annotation on the create method specifies text/xml as the response type. The method generates this response type with the XMLEncoder.

In addition to the read method, there is a second method, personalized_read, annotated with @GET. The method also has the annotation @Path("{extra}"). For example, the request:

```
% curl http://localhost:9876/bye
```

causes this method to be invoked with bye as the argument. The braces surrounding extra signal that extra is simply a placeholder rather than a literal. A method-level @Path is appended to the class-level @Path. In this example, the class-level @Path is simply /.

The Grizzly publisher automatically generates a WADL document, which is available at:

```
http://localhost:9876/application.wadl
```

Here is the automatically generated WADL for the MsgResource:

```xml
<?xml version="1.0" encoding="UTF-8" standalone="yes"?>
<application xmlns="http://research.sun.com/wadl/2006/10">
    <doc xmlns:jersey="http://jersey.dev.java.net/"
        jersey:generatedBy="Jersey: 0.9-ea 08/22/2008 04:48 PM"/>
    <resources base="http://localhost:9876/">
        <resource path="/">
            <method name="DELETE" id="delete">
                <response>
                    <representation mediaType="text/plain"/>
                </response>
            </method>
            <method name="GET" id="read">
                <response>
                    <representation mediaType="text/plain"/>
                </response>
            </method>
            <method name="POST" id="create">
                <request>
                    <param xmlns:xs="http://www.w3.org/2001/XMLSchema"
                        type="xs:string" name="msg"/>
                </request>
                <response>
                    <representation mediaType="text/xml"/>
                </response>
            </method>
            <resource path="{extra}">
                <param xmlns:xs="http://www.w3.org/2001/XMLSchema"
                    type="xs:string" style="template" name="extra"/>
                <method name="GET" id="personalized_read">
                    <response>
                        <representation mediaType="text/plain"/>
                    </response>
                </method>
            </resource>
```

```
        </resource>
      </resources>
    </application>
```

The WADL captures that the `MsgResource` supports two GET operations, a POST operation, and a DELETE operation. The WADL also describes the MIME type of the response representation for each operation. Of course, this WADL document could be used as input to the *wadl2java* utility.

Jersey is an appropriately lightweight framework that honors the spirit of RESTful services. The Grizzly publisher is attractive for development, automatically generating a WADL document to describe the published services. For production, the move to a web container, standalone or embedded in an application server, is straightforward because Jersey resources are simply annotated POJOs. The entire JSR-311 API, the Jersey core, comprises only 3 packages and roughly 50 interfaces and classes.

For now, JAX-WS and JAX-RS are separate frameworks. It would not be surprising if, in the future, the two frameworks merged.

The Restlet Framework

Several web frameworks have embraced REST, perhaps none more decisively than Rails with its `ActiveResource` type, which implements a resource in the RESTful sense. Rails also emphasizes a RESTful style in routing, with CRUD request operations specified as standard HTTP verbs. Grails is a Rails knockoff implemented in Groovy, which in turn is a Ruby knockoff with access to the standard Java packages. Apache Sling (*http://incubator.apache.org/sling/site/index.html*) is a Java-based web framework with a RESTful orientation.

The restlet framework (*http://www.restlet.org*) adheres to the REST architectural style and draws inspiration from other lightweight but powerful frameworks such as Net-Kernel (*http://www.1060.org*) and Rails. As the name indicates, a restlet is a RESTful alternative to the traditional Java servlet. The restlet framework has a client and a service API. The framework is well designed, relatively straightforward, professionally implemented, and well documented. It plays well with existing technologies. For example, a restlet can be deployed in a servlet container such as Tomcat or Jetty. The restlet distribution includes integration support for the Spring framework and also comes with the Simple HTTP engine (*http://www.simpleframework.org*), which can be embedded in Java applications. The sample restlet in this section is published with the Simple HTTP engine.

The `FibRestlet` application reprises the Fibonacci example yet again. The service, published with the Simple HTTP engine, illustrates key constructs in a restlet. Here is the source code:

```java
package ch04.restlet;

import java.util.Collections;
import java.util.Map;
import java.util.HashMap;
import java.util.Collection;
import java.util.List;
import java.util.ArrayList;
import org.restlet.Component;
import org.restlet.Restlet;
import org.restlet.data.Form;
import org.restlet.data.MediaType;
import org.restlet.data.Method;
import org.restlet.data.Parameter;
import org.restlet.data.Protocol;
import org.restlet.data.Request;
import org.restlet.data.Response;
import org.restlet.data.Status;

public class FibRestlet {
    private Map<Integer, Integer> cache =
        Collections.synchronizedMap(new HashMap<Integer, Integer>());
    private final String xml_start = "<fib:response xmlns:fib = 'urn:fib'>";
    private final String xml_stop = "</fib:response>";

    public static void main(String[ ] args) {
        new FibRestlet().publish_service();
    }

    private void publish_service() {
        try {
            // Create a component to deploy as a service.
            Component component = new Component();

            // Add an HTTP server to connect clients to the component.
            // In this case, the Simple HTTP engine is the server.
            component.getServers().add(Protocol.HTTP, 7777);

            // Attach a handler to handle client requests. (Note the
            // similarity of the handle method to an HttpServlet
            // method such as doGet or doPost.)
            Restlet handler = new Restlet(component.getContext()) {
                @Override
                public void handle(Request req, Response res) {
                    Method http_verb = req.getMethod();

                    if (http_verb.equals(Method.GET)) {
                        String xml = to_xml();
                        res.setStatus(Status.SUCCESS_OK);
                        res.setEntity(xml, MediaType.APPLICATION_XML);
                    }
                    else if (http_verb.equals(Method.POST)) {
                        // The HTTP form contains key/value pairs.
                        Form form = req.getEntityAsForm();
                        String nums = form.getFirstValue("nums");
```

```
                if (nums != null) {
                    // nums should be a list in the form: "[1, 2, 3]"
                    nums = nums.replace('[', '\0');
                    nums = nums.replace(']', '\0');
                    String[ ] parts = nums.split(",");
                    List<Integer> list = new ArrayList<Integer>();
                    for (String next : parts) {
                        int n = Integer.parseInt(next.trim());
                        cache.put(n, countRabbits(n));
                        list.add(cache.get(n));
                    }
                    String xml =
                      xml_start + "POSTed: " + list.toString() + xml_stop;
                    res.setStatus(Status.SUCCESS_OK);
                    res.setEntity(xml, MediaType.APPLICATION_XML);
                }
            }
            else if (http_verb.equals(Method.DELETE)) {
                cache.clear(); // remove the resource
                String xml =
                    xml_start + "Resource deleted" + xml_stop;
                res.setStatus(Status.SUCCESS_OK);
                res.setEntity(xml, MediaType.APPLICATION_XML);

            }
            else // only GET, POST, and DELETE supported
                res.setStatus(Status.SERVER_ERROR_NOT_IMPLEMENTED);
        }};

        // Publish the component as a service and start the service.
        System.out.println("FibRestlet at: http://localhost:7777/fib");
        component.getDefaultHost().attach("/fib", handler);
        component.start();
    }
    catch (Exception e) { System.err.println(e); }
}

private String to_xml() {
    Collection<Integer> list = cache.values();
    return xml_start + "GET: " + list.toString() + xml_stop;
}

private int countRabbits(int n) {
    n = Math.abs(n); // eliminate possibility of a negative argument

    // Easy cases.
    if (n < 2) return n;

    // Return cached values if present.
    if (cache.containsKey(n)) return cache.get(n);
    if (cache.containsKey(n - 1) &&
        cache.containsKey(n - 2)) {
      cache.put(n, cache.get(n - 1) + cache.get(n - 2));
      return cache.get(n);
    }
```

```
            // Otherwise, compute from scratch, cache, and return.
            int fib = 1, prev = 0;
            for (int i = 2; i <= n; i++) {
                int temp = fib;
                fib += prev;
                prev = temp;
            }
            cache.put(n, fib);
            return fib;
        }
    }
```

As the source code shows, the restlet framework provides easy-to-use Java wrappers such as `Method`, `Request`, `Response`, `Form`, `Status`, and `MediaType` for HTTP and MIME constructs. The framework supports virtual hosts for commercial-grade applications.

The restlet download includes a subdirectory *RESTLET_HOME/lib* that houses the various JAR files for the restlet framework itself and for interoperability with Tomcat, Jetty, Spring, Simple, and so forth. For the sample restlet service in this section, the JAR files *com.noelios.restlet.jar*, *org.restlet.jar*, and *org.simpleframework.jar* must be on the classpath.

A restlet client could be written using a standard class such as `HttpURLConnection`, of course. The following client illustrates the restlet API on the client side, an API that could be used independently of the service API:

```
import org.restlet.Client;
import org.restlet.data.Form;
import org.restlet.data.Method;
import org.restlet.data.Protocol;
import org.restlet.data.Request;
import org.restlet.data.Response;
import java.util.List;
import java.util.ArrayList;
import java.io.IOException;

class RestletClient {
    public static void main(String[ ] args) {
        new RestletClient().send_requests();
    }

    private void send_requests() {
        try {
            // Setup the request.
            Request request = new Request();
            request.setResourceRef("http://localhost:7777/fib");

            // To begin, a POST to create some service data.
            List<Integer> nums = new ArrayList<Integer>();
            for (int i = 0; i < 12; i++) nums.add(i);

            Form http_form = new Form();
            http_form.add("nums", nums.toString());
```

```
                request.setMethod(Method.POST);
                request.setEntity(http_form.getWebRepresentation());

                // Generate a client and make the call.
                Client client = new Client(Protocol.HTTP);

                // POST request
                Response response = get_response(client, request);
                dump(response);

                // GET request to confirm POST
                request.setMethod(Method.GET);
                request.setEntity(null);
                response = get_response(client, request);
                dump(response);

                // DELETE request
                request.setMethod(Method.DELETE);
                request.setEntity(null);
                response = get_response(client, request);
                dump(response);

                // GET request to confirm DELETE
                request.setMethod(Method.GET);
                request.setEntity(null);
                response = get_response(client, request);
                dump(response);
            }
            catch(Exception e) { System.err.println(e); }
        }

        private Response get_response(Client client, Request request) {
            return client.handle(request);
        }

        private void dump(Response response) {
            try {
                if (response.getStatus().isSuccess())
                    response.getEntity().write(System.out);
                else
                    System.err.println(response.getStatus().getDescription());
            }
            catch(IOException e) { System.err.println(e); }
        }
    }
```

The client API is remarkably clean. In this example, the client issues POST and DELETE
requests with GET requests to confirm that the *create* and *delete* operations against the
service were successful.

The restlet framework is a quick study. Its chief appeal is its RESTful orientation, which results in a lightweight but powerful software environment for developing and consuming RESTful services. The chief issue is whether the restlet framework can gain the market and mind share to become the standard environment for RESTful services in Java. The Jersey framework has the JSR seal of approval, which gives this framework a clear advantage.

What's Next?

Web services, whether SOAP based or REST style, likely require security. The term *security* is broad and vague. The next chapter clarifies the notion and explores the technologies available for securing web services. The emphasis is on user authentication and authorization, mutual challenge, and message encryption and decryption.

## CHAPTER 5

Web Services Security

Overview of Web Services Security

Web services security covers a lot of territory, which cannot be explored all at once. The territory is sufficiently broad that it needs to be divided into smaller, more manageable chunks. Here is a sketch of how this chapter and the next cover this territory:

Wire-level security
> Security begins at the transport or wire level; that is, with basic protocols that govern communications between a web service, whether SOAP-based or REST-style, and its clients. Security at this level typically provides three services. First, the client and service need transport-level assurance that each is communicating with the other rather than with some impostor. Second, the data sent from one side to the other needs to be encrypted strongly enough so that an interceptor cannot decrypt the data and thus gain access to the secrets carried therein. Third, each side needs assurance that the received message is the same as the sent message. This chapter covers the basics of wire-level security with code examples.

User authentication and authorization
> Web services provide clients with access to resources. If a resource is secured, then a client needs the appropriate credentials to gain access. The credentials are presented and verified through a process that usually has two phases. In the first phase, a client (user) presents information such as a username together with a credential such as a password. If the credential is not accepted, access to the requested resource is denied. The first phase is known as *user authentication*. The second phase, which is optional, consists of fine-tuning the authenticated user's access rights. For example, a stock-picking web service might provide all paying customers with a username and password, but the service might divide the customers into categories, for instance, *regular* and *premier*. Access to certain resources might be restricted to *premier* clients. The second phase is known as *role authorization*. This chapter introduces *users-role security*, a common name for the two-phase process, and the next chapter builds on the introduction.

WS-Security

WS-Security, or *WSS* for short, is a collection of protocols that specify how different levels of security can be enforced on messaging in SOAP-based web services. For example, WSS specifies how digital signatures and encryption information can be inserted into SOAP headers. Recall that SOAP-based services are designed to be transport-neutral. Accordingly, WSS is meant to provide comprehensive end-to-end security regardless of the underlying transport. This chapter introduces WS-Security with an example published with `Endpoint`, and the next chapter builds on the introduction by examining WSS in an application server such as GlassFish.

Wire-Level Security

Consider a pay-for web service such as Amazon's S3 storage service. This service needs to authenticate requests to store and retrieve data so that only the paying clients have access to the service and that, moreover, a particular client has privileged access to its paid-for storage. In the RESTful version of S3, Amazon uses a customization of *keyed HMAC* (Hash Message Authentication Code) to authenticate client requests. Here is a summary of how the authentication works:

- Parts of the request are concatenated together to form a single string, which becomes the input value for a hash computation. This string is the *input message*.

- The *AWS* (Amazon Web Services) secret access key, a unique bit string that Amazon provides to each paying client, is used to compute the hash value of the *input message* (see Figure 5-1). A hash value is also called a *message digest*, which is a fixed-length digest of arbitrarily many input bits. Amazon uses the *SHA-1* (Secure Hash Algorithm-1) version of HMAC, which produces a 160-bit digest no matter what the bit length of the input may be. Amazon calls this hash value the *signature* because the value functions like a digital signature, although technically a digital signature is an *encrypted* message digest. What Amazon calls the *signature* is not encrypted.

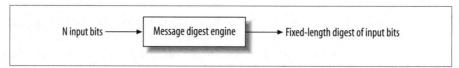

Figure 5-1. A message digest

- The Amazon signature is added to the request in the HTTP 1.1 `Authorization` header.

- Upon receipt of the request, the Amazon S3 first validates the Amazon signature and then honors the request only if the validation succeeds.

What prevents a client's request to Amazon S3 from being intercepted and the value of its `Authorization` header, the Amazon authentication signature, from being pirated?

Amazon assumes that the request is sent over the secure communications channel that *HTTPS* (HyperText Transport Protocol over Secure Socket Layer) provides. HTTPS is HTTP with an added security layer. Netscape did the original work in the design and implementation of this security layer and called it *SSL* (Secure Sockets Layer). The *IETF* (International Engineering Task Force) has taken over SSL and renamed it to *TLS* (Transport Layer Security). Although SSL and TLS differ in version numbers and in a few technical details, they are fundamentally the same. It is therefore common to use SSL, TLS, and SSL/TLS interchangeably.

Java has various packages that support SSL/TLS in general and HTTPS in particular. *JSSE* (Java Secure Sockets Extension) has been part of core Java since JDK 1.4. Of interest here is that higher levels of security, such as user authentication, usually require wire-level security of the kind that HTTPS provides. Accordingly, the discussion of web services security begins with HTTPS.

HTTPS Basics

HTTPS is easily the most popular among the secure versions of HTTP. HTTPS provides three critical security services over and above the transport services that HTTP provides. Following is a summary of the three, with Figure 5-2 as a reference. In the figure, Alice needs to send a secret message to Bob. Eve, however, may be eavesdropping. Eve may even try to dupe Alice and Bob into believing that they are communicating with one another when, in fact, each is communicating instead with Eve. This variation is known as the *MITM* (Man In The Middle) attack. For secure communications, Alice and Bob thus need these three services:

Peer authentication
> Alice needs Bob to authenticate himself so that she is sure about who is on the receiving end before she sends the secret message. Bob, too, needs Alice to authenticate herself so that he knows that the secret message is from her rather than an impostor such as Eve. This step also is described as *mutual authentication* or *mutual challenge*.

Confidentiality
> Once Alice and Bob have authenticated each other, Alice needs to encrypt the secret message in such a way that only Bob can decrypt it. Even if Eve intercepts the encrypted message, she should not be able to decrypt the message because doing so requires enormous computational power or incredibly good luck.

Integrity
> The message that Alice sends should be identical to the one that Bob receives. If not, an error condition should be raised. The received message might differ from the sent one for various reasons; for instance, noise in the communications channel or deliberate tampering by Eve. Any difference between the sent and the received message should be detected.

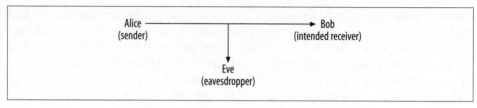

Figure 5-2. A secret message from Bob to Alice despite Eve

These features can be implemented in different ways, of course. Before considering how HTTPS implements the three features, it will be useful to look briefly at data encryption and decryption.

Symmetric and Asymmetric Encryption/Decryption

Modern approaches to encryption follow two different approaches, symmetric and asymmetric. Under either approach, the bits to be encrypted (*plain bits*) are one input to an encryption engine and an encryption *key* is the other input (see Figure 5-3). The encrypted bits are the *cipher bits*. If the input bits represent text, then they are the *plaintext* and the output bits are the *ciphertext*. The cipher bits are one input to the decryption engine, and a decryption key is the other input. The decryption produces the original plain bits. In the symmetric approach, the *same* key—called the *secret key* or the *single key*—is used to encrypt and decrypt (see Figure 5-4). The symmetric approach has the advantage of being relatively fast, but the disadvantage of what is known as the key distribution problem. How is the secret key itself to be distributed to the sender and the receiver?

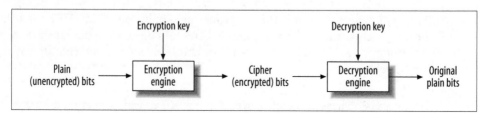

Figure 5-3. Basic encryption and decryption

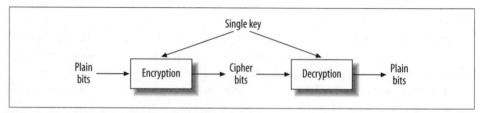

Figure 5-4. Single key encryption and decryption

In the asymmetric approach, the starting point is a *key pair*, which consists of a *private key* and a *public key*. As the names suggest, the private key should not be distributed but safeguarded by whoever generated the key pair. The public key can be distributed freely and publicly. If message bits are encrypted with the public key, they can be decrypted only with the private key, and vice-versa. Figure 5-5 illustrates. The asymmetric approach solves the key distribution problem, but asymmetric encryption and decryption are roughly a thousand times slower than their symmetric counterparts.

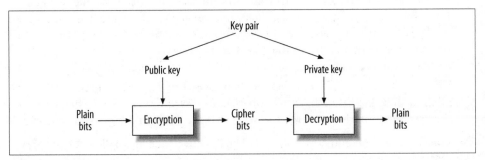

Figure 5-5. Public key encryption and decryption

The public key approach solves the confidentiality problem for Alice and Bob. If Alice encrypts the message with the public key from Bob's key pair, and Bob has the only copy of the private key from this pair, then only Bob can decrypt the message. Even if Eve intercepts Alice's message, she cannot decrypt it with Bob's public key.

How HTTPS Provides the Three Security Services

Of the three required security services—peer authentication, confidentiality, and integrity—the last is the most straightforward. The message sent over HTTPS includes a digest, which the receiver recomputes. If the sent digest differs from the digest that the receiver computes, then the message was altered during transmission, either by accident or design. If the sent digest itself is altered during transmission, this likewise counts as integrity failure.

HTTPS handles peer authentication through the exchange of digital certificates. In many cases, however, it is only the client that challenges the server. Consider a typical web application in which a shopper finalizes an order for the contents of his shopping cart by submitting a credit card number to the vendor. Here is a summary of what typically happens when the client-side browser and the web server negotiate to set up an HTTPS connection:

- The customer's browser challenges the vendor's web server to authenticate itself, and the server responds by sending one or more digital certificates to the browser.
- The browser checks the web server's digital certificates against the browser's *trust-store*, which is a database of digital certificates that the browser trusts. The

browser's validation of an incoming certificate can be and, for practical reasons, typically is indirect. For example, suppose that the browser receives a certificate from Amazon but does not have such a certificate in its truststore. Suppose further that the Amazon certificate contains a vouching signature from VeriSign, a well-known certificate authority (CA). If the browser's truststore has a VeriSign certificate, then the browser can use the VeriSign certificate to validate the VeriSign signature on the Amazon certificate. The point of interest is that the browser's truststore is its repository of certificates that are used to verify incoming certificates. If the browser cannot validate an incoming certificate against its truststore, then the browser typically asks the user whether the certificate should be trusted this time only or permanently. If the user selects *permanently*, the browser adds the certificate to its truststore.

- The web server typically does *not* challenge the browser. For one thing, the web application is interested in the shopper's credit card rather than in the identity of the user agent (in this case, the browser) that the shopper happens to be using.

The usually one-sided authentication challenge, with the client challenging the server but not the other way around, shows up in Tomcat's configuration file, *server.xml*. Here is the entry for HTTPS:

```
<Connector port="8443" protocol="HTTP/1.1" SSLEnabled="true"
           maxThreads="150" scheme="https" secure="true"
           clientAuth="false" sslProtocol="TLS" />
```

The clientAuth attribute is set to false, thereby indicating that Tomcat does not challenge the client. The default client behavior is to challenge the server.

For authentication and confidentiality, HTTPS relies on digital certificates, which are widely used in public key cryptography precisely because the exchange of secret keys is so difficult among many users. Here is a summary of how HTTPS authentication is intertwined with HTTPS confidentiality. The process is sometimes described as the *handshake* between client and server that culminates in a secure network connection. In this scenario, the client might be a browser or an application functioning as a web service client. For convenience, the term *web server* covers both a standard web application server such as Tomcat, an Endpoint publisher of a web service, or a full-bodied Java Application Server such as BEA WebLogic, GlassFish, JBoss, or WebSphere:

- The client challenges the web server, which sends one or more digital certificates as authentication. Modern digital certificates usually follow the X.509 format. The current X.509 version is named v3.

- An X.509 certificate is a *public key certificate* that serves as an *identity certificate* by binding the public key from a key pair to an identity such as a person (for instance, Alice) or an organization (for instance, Bob's employer). The certificate contains the digital signature of a CA such as a VeriSign, although certificates can be self-signed for testing purposes. In signing a digital certificate, a CA endorses the certificate and thereby verifies that the certificate's public key is bound to a particular identity. For instance, VeriSign signs Alice's certificate and thereby verifies that the certificate's public key belongs to Alice's key pair.

- The client can determine whether to accept the server's digital certificates by checking these against its truststore.

- The server has the option of challenging the client.

- Once the challenge phase is over, the client begins the process of generating a secret key. The process begins with the client's generation of a *pre-master secret*, a string, that is shared with the server. The pre-master is then used on each side to generate the same *master* secret key, which is used to encrypt and decrypt the traffic between the client and the server. At issue here is how the pre-master secret is sent securely from the client to the server.

- The client encrypts the 48-bit pre-master secret with the server's public key, available on the server's digital certificate downloaded during the peer authentication phase. The encrypted pre-master secret is sent to the server, which decrypts the secret. If everything is still in order, each side confirms that encryption of traffic between them is to begin. The public key/private key pair is thus critical in solving the key distribution problem for secret keys.

- At any point, either the client or the server may insist on starting the entire process all over again. For example, if either Alice or Bob suspects that Eve is up to no good, either Alice or Bob can restart the handshake process.

A secret key is used to encrypt and decrypt traffic for several reasons. First, symmetric encryption has relatively high performance. Second, if the server does not challenge the client, then the server does not have the client's public key to encrypt messages to the client. The server cannot encrypt messages with its own private key, as any receiver (for instance, Eve) with access to the server's public key then could decrypt the message. Finally, encrypting and decrypting with two separate key pairs is inherently trickier than using a single secret key.

The challenge is to get the pre-master secret securely from the client to the server; and the server's public key, available to the client in the server's digital certificate after the mutual challenge phase, fits the bill perfectly. The master secret key is generated only after the client and the server have agreed upon which *cipher suite*, or set of cryptographic algorithms, should be used. A cipher suite includes a key-pair algorithm and a hash algorithm.

Although digital certificates now play a dominant role in mutual challenge scenarios, they are not the only game in town. For example, *SRP* (Secure Remote Protocol) implements mutual challenge but without digital certificates. For more on SRP, see *http://srp.stanford.edu*.

The HttpsURLConnection Class

It is time to flesh out these architectural sketches with a code example. The HttpsURLConnection class, which extends HttpURLConnection, supports HTTPS connections. The SunClient application shown below uses an HttpsURLConnection to issue a GET request under HTTPS against the Java home site. Note that the port number in the URL is 443, the standard port for HTTPS connections:

```java
import java.net.URL;
import javax.net.ssl.HttpsURLConnection;
import java.net.MalformedURLException;
import java.security.cert.Certificate;
import java.io.IOException;
import java.io.BufferedReader;
import java.io.InputStreamReader;

// Send a GET request over HTTPS to the Java home site and
// display information about the security artifacts that come back.
class SunClient {
    private static final String url_s = "https://java.sun.com:443";

    public static void main(String[ ] args) {
        new SunClient().do_it();
    }
    private void do_it() {
        try {
            URL url = new URL(url_s);
            HttpsURLConnection conn = (HttpsURLConnection) url.openConnection();
            conn.setDoInput(true);
            conn.setRequestMethod("GET");
            conn.connect();
            dump_features(conn);
        }
        catch(MalformedURLException e) { System.err.println(e); }
        catch(IOException e) { System.err.println(e); }
    }
    private void dump_features(HttpsURLConnection conn) {
        try {
            print("Status code:   " + conn.getResponseCode());
            print("Cipher suite: " + conn.getCipherSuite());
            Certificate[ ] certs = conn.getServerCertificates();
            for (Certificate cert : certs) {
                print("\tCert. type: " + cert.getType());
                print("\tHash code:   " + cert.hashCode());
                print("\tAlgorithm:   " + cert.getPublicKey().getAlgorithm());
                print("\tFormat:      " + cert.getPublicKey().getFormat());
                print("");
            }
        }
        catch(Exception e) { System.err.println(e); }
    }
    private void print(String s) { System.out.println(s); }
}
```

Here is the output from a sample run:

```
Status code:  200
Cipher suite: SSL_RSA_WITH_RC4_128_MD5
        Cert. type: X.509
        Hash code:  23073427
        Algorithm:  RSA
        Format:     X.509

        Cert. type: X.509
        Hash code:  32560810
        Algorithm:  RSA
        Format:     X.509

        Cert. type: X.509
        Hash code:  8222443
        Algorithm:  RSA
        Format:     X.509
```

The status code of 200 signals that the GET request was successful. The SSL cipher suite can be read as follows:

RSA

This is the *public key cryptography algorithm*, named after Rivest, Shamir, and Adleman, the former MIT professors who designed it. RSA is the most commonly used public key algorithm. It is used to encrypt the pre-master that is sent from the client to the server.

RC4_128

The *stream cipher algorithm*, which is used to encrypt and decrypt the bit traffic between client and server, has a key length of 128 bits. The R is for Rivest in RSA, and the C is for cipher. (Sometimes RC is said to be shorthand for *Ron's Code*, as Rivest's first name is Ron.) RC4 is the most commonly used stream cipher. RC4_128 is used to encrypt the data traffic once the handshake is completed.

MD5

The certificate's 128-bit identifying hash, also called its *fingerprint*, is generated with the Message Digest algorithm 5. MD5 supplies what is officially known as the cryptographic hash function. Rivest designed MD5 as an improvement over MD4. Although MD5 is not fatally flawed, it is gradually losing ground to alternatives such as the Secure Hash Algorithm (SHA) family of hash algorithms.

The Sun server sent three digital certificates during the mutual challenge phase. Each is an X.509 certificate generated with the RSA algorithm. Each of the distinct MD5 digests is 128 bits in length. The format of each X.509 certificate understandably follows the X.509 specification.

The three X.509 digital certificates sent to the SunClient application can be validated against the client's default truststore, which is *JAVA_HOME/jre/lib/security/cacerts*. In development phase, however, it may be useful to turn off validation because, for example, the truststore has does not contain the appropriate certificates.

The SunTrustingClient application below revises the SunClient application by turning off certificate validation:

```java
import java.net.URL;
import java.security.SecureRandom;
import java.security.cert.X509Certificate;
import javax.net.ssl.SSLContext;
import javax.net.ssl.HttpsURLConnection;
import javax.net.ssl.TrustManager;
import javax.net.ssl.X509TrustManager;
import java.net.MalformedURLException;
import java.security.cert.Certificate;
import java.io.IOException;
import java.io.BufferedReader;
import java.io.InputStreamReader;

class SunTrustingClient {
    private static final String url_s = "https://java.sun.com:443";

    // Send a GET request and print the response status code.
    public static void main(String[ ] args) {
        new SunTrustingClient().do_it();
    }
    private void do_it() {
        try {
            // Configure HttpsURLConnection so that it doesn't check certificates.
            SSLContext ssl_ctx = SSLContext.getInstance("SSL");
            TrustManager[ ] trust_mgr = get_trust_mgr();
            ssl_ctx.init(null,              // key manager
                         trust_mgr,         // trust manager
                         new SecureRandom()); // random number generator
            HttpsURLConnection.setDefaultSSLSocketFactory(ssl_ctx.getSocketFactory());
            URL url = new URL(url_s);
            HttpsURLConnection conn = (HttpsURLConnection) url.openConnection();
            conn.setDoInput(true);
            conn.setRequestMethod("GET");
            conn.connect();
            dump_features(conn);
        }
        catch(MalformedURLException e) { System.err.println(e); }
        catch(IOException e) { System.err.println(e); }
        catch(Exception e) { System.err.println(e); }
    }
    private TrustManager[ ] get_trust_mgr() {
        // No exceptions thrown in the three overridden methods.
        TrustManager[ ] certs = new TrustManager[ ] {
          new X509TrustManager() {
            public X509Certificate[ ] getAcceptedIssuers() { return null; }
            public void checkClientTrusted(X509Certificate[ ] c, String t) { }
            public void checkServerTrusted(X509Certificate[ ] c, String t) { }
          }
        };
        return certs;
    }
```

```
private void dump_features(HttpsURLConnection conn) {
    try {
        print("Status code:  " + conn.getResponseCode());
        print("Cipher suite: " + conn.getCipherSuite());
        Certificate[ ] certs = conn.getServerCertificates();
        for (Certificate cert : certs) {
            print("\tCert. type: " + cert.getType());
            print("\tHash code:  " + cert.hashCode());
            print("\tAlgorithm:  " + cert.getPublicKey().getAlgorithm());
            print("\tFormat:     " + cert.getPublicKey().getFormat());
            print("");
        }

    }
    catch(Exception e) { System.err.println(e); }
}

private void print(String s) {
    System.out.println(s);
}
}
```

The revised application constructs its own `TrustManager` to override the default. A trust manager validates certificates. The output for the `SunClient` and the output for the `SunTrustingClient` is essentially the same, however, because even the trusting version still receives three certificates from the Sun server.

Securing the RabbitCounter Service

The `Endpoint.publish` method does not support HTTPS connections. However, core Java 6 includes an `HttpServer` class and its subclass `HttpsServer`, which can be used to publish a service under HTTPS. (The `Endpoint` publisher, in fact, uses the `HttpServer` class under the hood.) The `HttpsServer` can be used by itself to publish a RESTful service such as the `RabbitCounter`. Here is a revised publisher for the `RabbitCounter` service, a publisher that accepts HTTPS connections:

```
package ch05.rc;

import java.net.InetSocketAddress;
import javax.net.ssl.SSLContext;
import javax.net.ssl.SSLParameters;
import javax.net.ssl.SSLEngine;
import javax.net.ssl.TrustManager;
import javax.net.ssl.X509TrustManager;
import java.security.cert.X509Certificate;
import java.security.SecureRandom;
import java.security.KeyStore;
import javax.net.ssl.KeyManagerFactory;
import javax.net.ssl.TrustManagerFactory;
import java.io.FileInputStream;
import javax.xml.ws.http.HTTPException;
import java.io.OutputStream;
```

```
import java.io.InputStream;
import java.io.IOException;
import java.util.Collections;
import java.util.Map;
import java.util.HashMap;
import java.util.Collection;
import java.util.List;
import java.util.ArrayList;
import com.sun.net.httpserver.HttpContext;
import com.sun.net.httpserver.HttpHandler;
import com.sun.net.httpserver.HttpsServer;
import com.sun.net.httpserver.HttpsConfigurator;
import com.sun.net.httpserver.HttpExchange;
import com.sun.net.httpserver.HttpsParameters;

public class RabbitCounterPublisher {
    private Map<Integer, Integer> fibs;

    public RabbitCounterPublisher() {
        fibs = Collections.synchronizedMap(new HashMap<Integer, Integer>());
    }

    public static void main(String[ ] args) {
        new RabbitCounterPublisher().publish();
    }

    public Map<Integer, Integer> getMap() { return fibs; }

    private void publish() {
        int port = 9876;
        String ip = "https://localhost:";
        String path = "/fib";
        String url_string = ip + port + path;

        HttpsServer server = get_https_server(ip, port, path);
        HttpContext http_ctx = server.createContext(path);

        System.out.println("Publishing RabbitCounter at " + url_string);
        if (server != null) server.start();
        else System.err.println("Failed to start server. Exiting.");
    }

    private HttpsServer get_https_server(String ip, int port, String path) {
        HttpsServer server = null;
        try {
            InetSocketAddress inet = new InetSocketAddress(port);
            // 2nd arg = max number of client requests to queue
            server = HttpsServer.create(inet, 5);

            SSLContext ssl_ctx = SSLContext.getInstance("TLS");
            // password for keystore
            char[ ] password = "qubits".toCharArray();
            KeyStore ks = KeyStore.getInstance("JKS");
            FileInputStream fis = new FileInputStream("rc.keystore");
            ks.load(fis, password);
```

```
                KeyManagerFactory kmf = KeyManagerFactory.getInstance("SunX509");
                kmf.init(ks, password);
                TrustManagerFactory tmf = TrustManagerFactory.getInstance("SunX509");
                tmf.init(ks);
                ssl_ctx.init(kmf.getKeyManagers(), tmf.getTrustManagers(), null);

                // Create SSL engine and configure HTTPS to use it.
                final SSLEngine eng = ssl_ctx.createSSLEngine();
                server.setHttpsConfigurator(new HttpsConfigurator(ssl_ctx) {
                        public void configure(HttpsParameters parms) {
                                parms.setCipherSuites(eng.getEnabledCipherSuites());
                                parms.setProtocols(eng.getEnabledProtocols());
                        }
                    });
                server.setExecutor(null); // use default

                HttpContext http_context =
                    server.createContext(path, new MyHttpHandler(this));
            }
        catch(Exception e) { System.err.println(e); }
        return server;
    }
}

// The handle method is called on a particular request context,
// in this case on any request to the server that ends with /fib.
class MyHttpHandler implements HttpHandler {
    private RabbitCounterPublisher pub;
    public MyHttpHandler(RabbitCounterPublisher pub) { this.pub = pub; }

    public void handle(HttpExchange ex) {
        String verb = ex.getRequestMethod().toUpperCase();
        if (verb.equals("GET"))         doGet(ex);
        else if (verb.equals("POST"))    doPost(ex);
        else if (verb.equals("DELETE")) doDelete(ex);
        else throw new HTTPException(405);
    }

    private void respond_to_client(HttpExchange ex, String res) {
        try {
            ex.sendResponseHeaders(200, 0); // 0 means: arbitrarily many bytes
            OutputStream out = ex.getResponseBody();
            out.write(res.getBytes());
            out.flush();
            ex.close(); // closes all streams
        }
        catch(IOException e) { System.err.println(e); }
    }
    private void doGet(HttpExchange ex) {
        Map<Integer, Integer> fibs = pub.getMap();
        Collection<Integer> list = fibs.values();
        respond_to_client(ex, list.toString());
    }
```

```java
        private void doPost(HttpExchange ex) {
            Map<Integer, Integer> fibs = pub.getMap();
            fibs.clear(); // clear to create a new map
            try {
                InputStream in = ex.getRequestBody();
                byte[ ] raw_bytes = new byte[4096];
                in.read(raw_bytes);
                String nums = new String(raw_bytes);
                nums = nums.replace('[', '\0');
                nums = nums.replace(']', '\0');
                String[ ] parts = nums.split(",");
                List<Integer> list = new ArrayList<Integer>();
                for (String next : parts) {
                    int n = Integer.parseInt(next.trim());
                    fibs.put(n, countRabbits(n));
                    list.add(fibs.get(n));
                }
                Collection<Integer> coll = fibs.values();
                String res = "POSTed: " + coll.toString();
                respond_to_client(ex, res);
            }
            catch(IOException e) { }
        }
        private void doDelete(HttpExchange ex) {
            Map<Integer, Integer> fibs = pub.getMap();
            fibs.clear();
            respond_to_client(ex, "All entries deleted.");
        }
        private int countRabbits(int n) {
            n = Math.abs(n);
            if (n < 2) return n; // easy cases

            Map<Integer, Integer> fibs = pub.getMap();
            // Return cached values if present.
            if (fibs.containsKey(n)) return fibs.get(n);
            if (fibs.containsKey(n - 1) &&
                fibs.containsKey(n - 2)) {
              fibs.put(n, fibs.get(n - 1) + fibs.get(n - 2));
              return fibs.get(n);
            }

            // Otherwise, compute from scratch, cache, and return.
            int fib = 1, prev = 0;
            for (int i = 2; i <= n; i++) {
                int temp = fib;
                fib += prev;
                prev = temp;
            }
            fibs.put(n, fib);
            return fib;
        }
    }
```

Creating an `HttpsServer` instance is straightforward:

```
HttpsServer server = null;
try {
    InetSocketAddress inet = new InetSocketAddress(port);
    server = HttpsServer.create(inet, 5);
```

The second argument to `create`, in this case 5, is the maximum number of requests that should be queued. A value of 0 signals that the system default should be used.

The code to configure an `HttpsServer` is trickier than the code to create one. There are several important features of the configuration, which centers on an `SSLContext` that is associated with a specific protocol such as TLS. In this example, the `SSLContext` is used for two configurations. The first configuration manages the keystore and truststore. Recall that the *keystore*, on the server side, contains the server's digital certificates that can be sent to the client during peer authentication; on the client side, the keystore contains the client's digital certificates for the same purpose. There is no default keystore. Here is the command used to generate the keystore file *rc.keystore* (the name is arbitrary):

```
% keytool -genkey -keyalg RSA -keystore rc.keystore
```

The command creates a single self-signed X.509 certificate and places this certificate in the file *rc.keystore*. The *keytool* utility, which comes with core Java 6, prompts for information that is used to build the X.509 certificate. The utility also asks for a password, in this case `qubits`. At runtime the keystore's contents will be loaded into the application so that, during the TLS handshake, any digital certificates therein can be sent to the client.

Next, the *truststore*, the database of trusted digital certificates, is configured. (Recall that there is a default truststore: *JAVA_HOME/jre/lib/security/cacerts*.) Once the keystore and truststore configuration is ready, the statement:

```
ssl_ctx.init(kmf.getKeyManagers(), tmf.getTrustManagers(), null);
```

makes these configurations part of the `SSLContext`. The third argument above, `null`, signals that the system should use the default `SecureRandom` number generator. The keystore and truststore configuration also could be done through `-D` command-line options or by setting various system properties in the code.

The last configuration step is to use the `SSLContext` to get an `SSLEngine`, which in turn provides the protocols and cipher suites available for establishing a secure connection. In this example, the system defaults are taken and no additional protocols or cipher suites are added to the already available mix.

The last setup step before starting the server is to establish an `HttpContext`, which consists of a path (in this case, `/fibs`) and an `HttpHandler` whose `handle` method is called back on every request for the resource at `/fibs`. The handler is associated with the `HttpsServer`. In effect, the HTTPS server delegates request processing to the specified handle method. An elaborate `HttpsServer` might have many `HttpContext` instances and separate handlers for each.

Here is the `MyHttpHandler`'s implementation of the `HttpHandler` method `handle`:

```java
public void handle(HttpExchange ex) {
    String verb = ex.getRequestMethod().toUpperCase();
    if (verb.equals("GET"))          doGet(ex);
    else if (verb.equals("POST"))    doPost(ex);
    else if (verb.equals("DELETE")) doDelete(ex);
    else throw new HTTPException(405); // bad verb
}
```

The `HttpExchange` argument encapsulates methods to access an input stream, which can be used to read client POST requests, and an output stream, which can be used to generate the body of a response to the client. There are methods to set the response's status code and the like. For example, here is the `respond_to_client` method that handles responses to GET, POST, and DELETE requests:

```java
private void respond_to_client(HttpExchange ex, String res) {
    try {
        ex.sendResponseHeaders(200, 0); // 0 means: arbitrarily many bytes
        OutputStream out = ex.getResponseBody();
        out.write(res.getBytes());
        out.flush();
        ex.close(); // closes all streams
    }
    catch(IOException e) { System.err.println(e); }
}
```

The HTTPS client against the HTTPS `RabbitCounter` service differs only slightly from the `SunClient` and the `SunTrustingClient` shown earlier. An `HttpsURLConnection` is used again and, for simplicity, is configured not to check the digital certificates sent from the server. The one change involves a `HostnameVerifier`, which allows the client to decide explicitly about whether it is all right to set up an HTTPS connection with a specified server. Here is the code segment, with `conn` as the reference to the `HttpsURLConnection` object:

```java
conn.setHostnameVerifier(new HostnameVerifier() {
        public boolean verify(String host, SSLSession sess) {
            if (host.equals("localhost")) return true;
            else return false;
        }});
```

The `verify` method can be coded to handle whatever connection logic is appropriate to the application. In this case, HTTPS connections to a server running on `localhost` are allowed, but attempted connections to any other server result in an exception. Following is the source code for the client, which sends a test POST request to the server:

```java
import java.util.List;
import java.util.ArrayList;
import java.net.URL;
```

```
import java.security.SecureRandom;
import java.security.cert.X509Certificate;
import javax.net.ssl.SSLContext;
import javax.net.ssl.HttpsURLConnection;
import javax.net.ssl.HostnameVerifier;
import javax.net.ssl.SSLSession;
import java.net.HttpURLConnection;
import javax.net.ssl.TrustManager;
import javax.net.ssl.X509TrustManager;
import java.net.MalformedURLException;
import java.security.cert.Certificate;
import java.io.InputStream;
import java.io.OutputStream;
import java.io.IOException;
import java.io.BufferedReader;
import java.io.InputStreamReader;

class SecureClientRC {
    private static final String url_s = "https://localhost:9876/fib";

    public static void main(String[ ] args) {
        new SecureClientRC().do_it();
    }

    private void do_it() {
        try {
            // Create a context that doesn't check certificates.
            SSLContext ssl_ctx = SSLContext.getInstance("TLS");
            TrustManager[ ] trust_mgr = get_trust_mgr();
            ssl_ctx.init(null,                    // key manager
                         trust_mgr,               // trust manager
                         new SecureRandom()); // random number generator
            HttpsURLConnection.setDefaultSSLSocketFactory(ssl_ctx.getSocketFactory());

            URL url = new URL(url_s);
            HttpsURLConnection conn = (HttpsURLConnection) url.openConnection();

            // Guard against "bad hostname" errors during handshake.
            conn.setHostnameVerifier(new HostnameVerifier() {
                    public boolean verify(String host, SSLSession sess) {
                        if (host.equals("localhost")) return true;
                        else return false;
                    }
                });

            // Test request
            List<Integer> nums = new ArrayList<Integer>();
            nums.add(3); nums.add(5); nums.add(7);

            conn.setDoInput(true);
            conn.setDoOutput(true);
            conn.setRequestMethod("POST");
            conn.connect();
            OutputStream out = conn.getOutputStream();
            out.write(nums.toString().getBytes());
```

```
            byte[ ] buffer = new byte[4096];
            InputStream in = conn.getInputStream();
            in.read(buffer);
            System.out.println(new String(buffer));
            dump_features(conn);
            conn.disconnect();
        }
        catch(MalformedURLException e) { System.err.println(e); }
        catch(IOException e) { System.err.println(e); }
        catch(Exception e) { System.err.println(e); }
    }

    private TrustManager[ ] get_trust_mgr() {
        TrustManager[ ] certs = new TrustManager[ ] {
            new X509TrustManager() {
                public X509Certificate[ ] getAcceptedIssuers() { return null; }
                public void checkClientTrusted(X509Certificate[ ] certs, String t) { }
                public void checkServerTrusted(X509Certificate[ ] certs, String t) { }
            }
        };
        return certs;
    }

    private void dump_features(HttpsURLConnection conn) {
        try {
            print("Status code:  " + conn.getResponseCode());
            print("Cipher suite: " + conn.getCipherSuite());
            Certificate[ ] certs = conn.getServerCertificates();
            for (Certificate cert : certs) {
                print("\tCert. type: " + cert.getType());
                print("\tHash code:  " + cert.hashCode());
                print("\tAlgorithm:  " + cert.getPublicKey().getAlgorithm());
                print("\tFormat:     " + cert.getPublicKey().getFormat());
                print("");
            }
        }
        catch(Exception e) { System.err.println(e); }
    }

    private void print(String s) { System.out.println(s); }
}
```

Here is the output:

```
POSTed: [2, 5, 13]
Status code:  200
Cipher suite: SSL_RSA_WITH_RC4_128_MD5
        Cert. type: X.509
        Hash code:  1992213
        Algorithm:  RSA
        Format:     X.509
```

The server's keystore, the file *rc.keystore*, contains just one self-signed X.509 certificate, but the cipher suite is the same as in the two Sun examples shown earlier. The *keytool* utility allows other cipher suites to be specified.

As proof of concept, here is a Perl client invoked against the `RabbitCounter` service after the Java client has posted the values shown previously:

```perl
#!/usr/bin/perl -w

use Net::SSLeay qw(get_https);
use strict;

my ($type, $start_line, $misc, $extra) = get_https('localhost', 9876, '/fib');
print "Type/value: $type\n";
print "Start line: $start_line\n";
print "Misc:       $misc => $extra\n";
```

By default, the Perl client does not verify the server's certificates, although this feature can be turned on easily. The Perl function `get_https` returns a list of four items.

The output was:

```
Type/value: a [2, 5, 13]

Start line: HTTP/1.1 200 OK
Misc:       TRANSFER-ENCODING => chunked
```

In the first line, the character `a` indicates that the service returns an array.

Adding User Authentication

To begin, here is a quick review of the jargon introduced earlier. User authentication is the first phase in a two-phase process known popularly as *users-roles security*. In the first phase, a user presents credentials such as a password in order to become an *authenticated subject*. More sophisticated credentials such as a smart card or biometric data (from, for instance, a fingerprint or even a retinal scan) might be used in user authentication.

The second phase, *authorization*, is optional and normally occurs only if the first phase succeeds. This phase determines which authorization roles an authenticated subject may play. Authorization roles can be customized as needed. For instance, Unix-based operating systems distinguish between *users* and *superusers*, which are authorization roles that determine levels of access to system resources and which privileged actions accompany a role. An IT organization might have authorization roles such as *programmer*, *senior database administrator*, *software engineer*, *systems analyst*, and so on. In large organizations, digital certificates may be used to determine authorization roles. For example, a system based on users-roles security may require Fred Flintstone to provide his username and password to become an authenticated subject but then check a database for a certificate that authorizes Fred as a crane operator.

HTTP BASIC Authentication

With a few minor changes, the RabbitCounter service can support what is known as HTTP BASIC Authentication. JAX-WS provides two convenient constants to look up usernames and passwords. The changes are local to the HttpHandler implementation. The handle method now invokes the authenticate method, which throws an HTTP 401 (Unauthorized) exception if the authentication fails and otherwise does nothing. See Example 5-1.

Example 5-1. Adding basic authentication to the RabbitCounter service

```
public void handle(HttpExchange ex) {
    authenticate(ex);
    String verb = ex.getRequestMethod().toUpperCase();
    if (verb.equals("GET"))         doGet(ex);
    else if (verb.equals("POST"))    doPost(ex);
    else if (verb.equals("DELETE")) doDelete(ex);
    else throw new HTTPException(405);
}

private void authenticate(HttpExchange ex) {
    // Extract the header entries.
    Headers headers = ex.getRequestHeaders();
    List<String> ulist = headers.get(BindingProvider.USERNAME_PROPERTY);
    List<String> plist = headers.get(BindingProvider.PASSWORD_PROPERTY);

    // Extract username/password from the two singleton lists.
    String username = ulist.get(0);
    String password = plist.get(0);
    if (!username.equals("fred") || !password.equals("rockbed"))
        throw new HTTPException(401); // authentication error
}
```

In production mode, of course, the username and password might be checked against a database. This change to the source also requires two additional import statements, one for BindingProvider and another for Headers.

On the client side, the change is also minor. The username and password are added to the HTTP request before the connect method is invoked on the HttpsURLConnection. In this code segment, conn is the reference to the HttpsURLConnection instance:

```
conn.addRequestProperty(BindingProvider.USERNAME_PROPERTY, "fred");
conn.addRequestProperty(BindingProvider.PASSWORD_PROPERTY, "rockbed");
```

The key point is that the authentication information is sent to the RabbitCounter service with wire-level security, in particular data encryption.

Container-Managed Security for Web Services

This most recent version of the RabbitCounter service does provide user authentication, but not in a way that scales. A better approach would be to use a web service container that provides not only user authentication but also wire-level security. Tomcat, the reference implementation for a Java web container, can provide both. Chapter 4 showed how Tomcat can be used to publish RESTful web services as servlets. Tomcat also can publish SOAP-based services. Tomcat can publish either a `@WebService` or a `@WebServiceProvider`.

The example to illustrate how Tomcat provides container-managed security is built in two steps. The first step publishes a SOAP-based service with Tomcat, and the second step adds security. A later example is a secured `@WebServiceProvider` under Tomcat.

Deploying a @WebService Under Tomcat

The SOAP-based service is organized in the usual way. Here is the code for the `TempConvert` SEI:

```
package ch05.tc;
import javax.jws.WebService;
import javax.jws.WebMethod;

@WebService
public interface TempConvert {
    @WebMethod float c2f(float c);
    @WebMethod float f2c(float f);
}
```

And here is the code for the corresponding SIB:

```
package ch05.tc;
import javax.jws.WebService;

@WebService(endpointInterface = "ch05.tc.TempConvert")
public class TempConvertImpl implements TempConvert {
    public float c2f(float t) { return 32.0f + (t * 9.0f / 5.0f); }
    public float f2c(float t) { return (5.0f / 9.0f) * (t - 32.0f); }
}
```

For deployment under Tomcat, the service requires a deployment descriptor, the *web.xml* file. This configuration file informs the servlet container about how requests for the service are to be handled. In particular, the *web.xml* refers to two special classes that come with JAX-WS 2.1: `WSServlet` and `WSServletContainerListener`. This segment of the *web.xml*:

```
<servlet>
   <servlet-name>TempConvertWS</servlet-name>
   <servlet-class>
      com.sun.xml.ws.transport.http.servlet.WSServlet
   </servlet-class>
</servlet>
<servlet-mapping>
   <servlet-name>TempConvertWS</servlet-name>
   <url-pattern>/tc</url-pattern>
</servlet-mapping>
```

delegates requests whose URLs end with the path /tc to a WSServlet instance, which in turn is linked to the JWS runtime. Tomcat provides the WSServlet instance.

The *web.xml* segment:

```
<listener>
  <listener-class>
     com.sun.xml.ws.transport.http.servlet.WSServletContextListener
  </listener-class>
</listener>
```

specifies a WSServletContextListener instance that will parse a second, Sun-specific configuration file named *sun-jaxws.xml*, which provides the web service's endpoint by connecting the WSServlet instance to the service's implementation class. Here is the *sun-jaxws.xml* file:

```
<?xml version="1.0" encoding="UTF-8"?>
<endpoints xmlns="http://java.sun.com/xml/ns/jax-ws/ri/runtime" version="2.0">
  <endpoint
     name="TempConvertWS"
     implementation="ch05.tc.TempConvertImpl"
     url-pattern="/tc"/>
</endpoints>
```

The name TempConvertWS is the name of the WSSerlvet in the *web.xml* file. As the syntax indicates, the configuration file could specify multiple service endpoints but, in this case, specifies only one.

In the deployed WAR file, the *web.xml* and *sun-jaxws.xml* files reside in the *WEB-INF* directory. For reference, here is the entire *web.xml* file:

```
<?xml version="1.0" encoding="UTF-8"?>
<web-app
     xmlns="http://java.sun.com/xml/ns/j2ee"
     xmlns:xsi="http://www.w3.org/2001/XMLSchema-instance"
     xsi:schemaLocation="http://java.sun.com/xml/ns/j2ee
                         http://java.sun.com/xml/ns/j2ee/web-app_2_4.xsd"
     version="2.4">
  <listener>
    <listener-class>
        com.sun.xml.ws.transport.http.servlet.WSServletContextListener
    </listener-class>
  </listener>
  <servlet>
    <servlet-name>TempConvertWS</servlet-name>
    <servlet-class>
        com.sun.xml.ws.transport.http.servlet.WSServlet
    </servlet-class>
  </servlet>
  <servlet-mapping>
    <servlet-name>TempConvertWS</servlet-name>
    <url-pattern>/tc</url-pattern>
  </servlet-mapping>
</web-app>
```

The service is deployed as if it were a standard Tomcat web application. Copies of the compiled SEI and SIB need to reside in the *WEB-INF/classes/ch05/tc* directory, and the two configuration files need to reside in the *WEB-INF* directory. Note that the service is, by default, `document-literal`; hence, the *wsgen* utility can be run:

```
% wsgen -cp . WEB-INF.classes.ch05.tc.TempConvertImpl
```

to generate the required JAX-B artifacts.

The WAR file, whose name is arbitrary, is then built:

```
% jar cvf tempc.war WEB-INF
```

and copied to the *TOMCAT_HOME/webapps* directory for deployment. The success of the deployment can be confirmed by checking a Tomcat log file or by opening a browser to the URL *http://localhost:8080/tempc/tc?wsdl*.

The *wsimport* utility now can be employed in the usual way to generate client-side artifacts from the WSDL:

```
% wsimport -keep -p tcClient http://localhost:8080/tempc/tc?wsdl
```

Here is a sample client, `ClientTC`, coded with the *wsimport* artifacts:

```
import tcClient.TempConvertImplService;
import tcClient.TempConvert;

class ClientTC {
    public static void main(String args[ ]) throws Exception {
        TempConvertImplService service = new TempConvertImplService();
        TempConvert port = service.getTempConvertImplPort();

        System.out.println("f2c(-40.1) ==> " + port.f2C(-40.1f));
        System.out.println("c2f(-40.1) ==> " + port.c2F(-40.1f));
        System.out.println("f2c(+98.7) ==> " + port.f2C(+98.7f));
    }
}
```

The output is:

```
f2c(-40.1) ==> -40.055557
c2f(-40.1) ==> -40.18
f2c(+98.7) ==> 37.055557
```

Securing the @WebService Under Tomcat

The `TempConvert` service does not need to be changed at all. This is the obvious benefit of having the container, Tomcat, rather than the application manage the security.

The first step is to ensure that the Tomcat *connector* for SSL/TLS is enabled. A connector is an endpoint for client requests. In the main Tomcat configuration file *conf/server.xml*, the section:

```
<Connector port="8443" protocol="HTTP/1.1" SSLEnabled="true" maxThreads="150"
           scheme="https" secure="true" clientAuth="false" sslProtocol="TLS" />
```

may have to be uncommented. If the section is commented out, then Tomcat has to be restarted after the editing change. By default, Tomcat awaits HTTPS requests on port 8443, although this port number can be changed.

When a client sends an HTTPS request to Tomcat, Tomcat will need a digital certificate to send back during the mutual challenge phase. The command:

```
% keytool -genkey -alias tomcat -keyalg RSA
```

generates a self-signed certificate. The *keytool* utility will prompt for various pieces of information, including a password, which should be set to changeit because this is what Tomcat expects. By default, the certificate is placed in a file named *.keystore* in the user's home directory, although the -keystore option can be used with the *keytool* utility to specify a different file name for the keystore.

On the local system, the user account that starts Tomcat needs to have a keystore file in the account's home directory. However, the Tomcat configuration file *server.xml* also can specify the location of the keystore file—and this seems the safest choice. Here is the amended entry for the HTTPS entry on my machine:

```
<Connector port="8443" protocol="HTTP/1.1" SSLEnabled="true" maxThreads="150"
           scheme="https" secure="true" clientAuth="false" sslProtocol="TLS"
           keystoreFile="/home/mkalin/.keystore" />
```

Tomcat now can be restarted and tested by entering the URL *https://localhost:8443* in a browser. Because the browser does not have the newly minted, self-signed certificate in its own truststore, the browser will prompt about whether to accept the certificate and, if so, whether to accept it permanently; that is, as an addition to the browser's truststore. Once the browser receives all of the required *yes* answers, it will display the Tomcat home page.

On the client side, the ClientTC application needs to be modified slightly. This client uses the *wsimport*-generated artifacts, in particular the class tcClient.TempConvertImpl Service, which has the endpoint URL set to *http://localhost:8080/tempc/tc?wsdl*. There is no reason to change the endpoint, as the WSDL document is still available at this URL. Yet the client needs to connect to *https://localhost:8443* rather than *http://localhost:8080*. The adjustment can be made in the client code using the BindingProvider. Here is the revised client:

```
import tcClient.TempConvertImplService;
import tcClient.TempConvert;
import javax.xml.ws.BindingProvider;
import java.util.Map;

class ClientTCSecure {
    private static final String endpoint = "https://localhost:8443/tempc/tc";
    public static void main(String args[ ]) {
        TempConvertImplService service = new TempConvertImplService();
        TempConvert port = service.getTempConvertImplPort();

        Map<String, Object> req_ctx = ((BindingProvider) port).getRequestContext();
        req_ctx.put(BindingProvider.ENDPOINT_ADDRESS_PROPERTY, endpoint);
```

```
        System.out.println("f2c(-40.1) ==> " + port.f2C(-40.1f));
        System.out.println("c2f(-40.1) ==> " + port.c2F(-40.1f));
        System.out.println("f2c(+98.7) ==> " + port.f2C(+98.7f));
    }
}
```

By default, Tomcat does not challenge the client during the peer authentication phase, but this could change. The client, in any case, cannot count on not being challenged and so must have a keystore with an identifying digital certificate. Further, the client needs a truststore against which to compare the certificate from Tomcat. Although the keystore and the truststore serve different purposes, each is a database of certificates with the same file format. To simplify the example, the Tomcat keystore will do triple duty by serving as the client keystore and truststore as well. The point of interest is the security architecture rather than the particular truststores and keystores. Here is the command to invoke the client, with the -D options to set the keystore and truststore information:

```
% java -Djavax.net.ssl.trustStore=/home/mkalin/.keystore \
       -Djavax.net.ssl.trustStorePassword=changeit       \
       -Djavax.net.ssl.keyStore=/home/mkalin/.keystore    \
       -Djavax.net.ssl.keyStorePassword=changeit ClientTC
```

The command is sufficiently complicated that an Ant or equivalent script might be used in production. The output is the same as before, of course.

Application-Managed Authentication

The next step in securing the TempConvert service is to add authentication and authorization. These additions can be done at the application level (that is, in the web service code itself) or at the container level by using shared resources such as a database of usernames and passwords. Container-managed authentication has the obvious appeal of letting Tomcat handle all of the security so that the service can focus on application logic. Container-managed authentication leads to cleaner code by adhering to the AOP practice of having the container—rather than the many different applications it contains—provide the security aspect.

Authentication at the application level does not involve much code but does muddy up the code with a mix of security concerns and business logic. Perhaps the most straightforward approach at the application level is to follow the idiom shown earlier in Example 5-1 by passing the username and password as part of the request context. Here is the revised ClientTC code:

```
import tcClient.TempConvertImplService;
import tcClient.TempConvert;
import javax.xml.ws.BindingProvider;
import javax.xml.ws.handler.MessageContext;
import java.util.Map;
import java.util.HashMap;
import java.util.Collections;
```

```
import java.util.List;

class ClientTC {
    private static final String endpoint = "https://localhost:8443/tempc/tc";

    public static void main(String args[ ]) {
        TempConvertImplService service = new TempConvertImplService();
        TempConvert port = service.getTempConvertImplPort();

        Map<String, Object> req_ctx = ((BindingProvider) port).getRequestContext();
        req_ctx.put(BindingProvider.ENDPOINT_ADDRESS_PROPERTY, endpoint);

        // Place the username/password in the HTTP request headers,
        // which a non-Java client can do as well.
        Map<String, List<String>> hdr = new HashMap<String, List<String>>();
        hdr.put("Username", Collections.singletonList("fred"));
        hdr.put("Password", Collections.singletonList("rockbed"));
        req_ctx.put(MessageContext.HTTP_REQUEST_HEADERS, hdr);

        // Invoke service methods.
        System.out.println("f2c(-40.1) ==> " + port.f2C(-40.1f));
        System.out.println("c2f(-40.1) ==> " + port.c2F(-40.1f));
        System.out.println("f2c(+98.7) ==> " + port.f2C(+98.7f));
    }
}
```

The revised ClientTC now places the username and password in the HTTP headers to underscore that there is nothing Java-specific in this approach. For clarity, the username and password are hardcoded, although they presumably would be entered as command-line arguments.

The change to the TempConvert service is also relatively minor. The additional method authenticated checks the username and password. In production, of course, the check would likely go against values stored in a database:

```
package ch05.tc;

import javax.jws.WebService;
import javax.annotation.Resource;
import javax.xml.ws.WebServiceContext;
import javax.xml.ws.handler.MessageContext;
import javax.xml.ws.http.HTTPException;
import java.util.Map;
import java.util.List;

@WebService(endpointInterface = "ch05.tc.TempConvert")
public class TempConvertImpl implements TempConvert {
    @Resource
    WebServiceContext ws_ctx;

    public float c2f(float t) {
        if (authenticated()) return 32.0f + (t * 9.0f / 5.0f);
        else throw new HTTPException(401); // authorization error
    }
```

```
    public float f2c(float t) {
        if (authenticated()) return (5.0f / 9.0f) * (t - 32.0f);
        else throw new HTTPException(401); // authorization error
    }

    private boolean authenticated() {
        MessageContext mctx = ws_ctx.getMessageContext();
        Map http_headers = (Map) mctx.get(MessageContext.HTTP_REQUEST_HEADERS);
        List ulist = (List) http_headers.get("Username");
        List plist = (List) http_headers.get("Password");

        // proof of concept authentication
        if (ulist.contains("fred") && plist.contains("rockbed")) return true;
        else return false;
    }
}
```

The downside of handling the authentication in the application code is evident. The web service operations, in this case f2c and c2f, are now a mix of application logic and security processing, however minor in this example. The authenticated method needs to access the MessageContext, the kind of low-level processing best left to handlers. The example is small enough that the mix is not overwhelming, but the example likewise suggests how complicated the intermix of application logic and security could become in a real-world service. Further, this approach does not scale well. If Fred needs to access other services that require authentication, then these services will have to replicate the kind of code shown here. An attractive option to the approach taken in this example is to let the container handle the authentication.

Container-Managed Authentication and Authorization

Tomcat provides container-managed authentication and authorization. The concept of a *realm* plays a central role in the Tomcat approach. A realm is a collection of resources, including web pages and web services, with a designated authentication and authorization facility. The Tomcat documentation describes a realm as being akin to a Unix group with respect to access rights. A realm is an organizational tool that allows a collection of resources to be under a single policy for access control.

Tomcat provides five standard plugins, with Realm in the names, to manage communications between the container and authentication/authorization store. Here are the five plugins with a short description of each:

JDBCRealm
> The authentication information is stored in a relational database accessible through a standard Java JDBC driver.

DataSourceRealm
> The authentication information again is stored in a relational database and accessible through a Java JDBC DataSource, which in turn is available through a *JNDI* (Java Naming and Directory Interface) lookup service.

JNDIRealm

> The authentication information is stored in an *LDAP*-based (Lightweight Directory Access Protocol) directory service, which is available through a JNDI provider.

MemoryRealm

> The authentication information is read into the container at startup from the file *conf/tomcat-users.xml*. This is the simplest choice and the default.

JAASRealm

> The authentication information is available through a *JAAS* (Java Authentication and Authorization Service) provider, which in turn is available in a Java Application Server such as BEA WebLogic, GlassFish, JBoss, or WebSphere.

Under any of these choices, it is the Tomcat container rather than the application that becomes the security provider.

Configuring Container-Managed Security Under Tomcat

The Tomcat approach to security also is *declarative* rather than *programmatic*; that is, details about the security realm are specified in a configuration file rather than in code. The configuration file is the *web.xml* document included in the deployed WAR file.

Here is the *web.xml* for the next revision to the `TempConvert` service, which uses the Tomcat MemoryRealm:

```
<?xml version="1.0" encoding="UTF-8"?>
<web-app xmlns="http://java.sun.com/xml/ns/j2ee"
         xmlns:xsi="http://www.w3.org/2001/XMLSchema-instance"
         xsi:schemaLocation="http://java.sun.com/xml/ns/j2ee
                     http://java.sun.com/xml/ns/j2ee/web-app_2_4.xsd" version="2.4">
    <listener>
      <listener-class>
        com.sun.xml.ws.transport.http.servlet.WSServletContextListener
      </listener-class>
    </listener>
    <servlet>
      <servlet-name>TempConvertWS</servlet-name>
      <servlet-class>
        com.sun.xml.ws.transport.http.servlet.WSServlet
      </servlet-class>
    </servlet>

    <security-role>
      <description>The Only Secure Role</description>
      <role-name>bigshot</role-name>
    </security-role>

    <security-constraint>
      <web-resource-collection>
        <web-resource-name>Users-Roles Security</web-resource-name>
        <url-pattern>/tcauth</url-pattern>
      </web-resource-collection>
```

```
    <auth-constraint>
      <role-name>bigshot</role-name>
    </auth-constraint>
    <user-data-constraint>
        <transport-guarantee>CONFIDENTIAL</transport-guarantee>
    </user-data-constraint>
  </security-constraint>

  <login-config>
     <auth-method>BASIC</auth-method>
  </login-config>

  <servlet-mapping>
    <servlet-name>TempConvertWS</servlet-name>
    <url-pattern>/tcauth</url-pattern>
  </servlet-mapping>
</web-app>
```

In the revised *web.xml*, there are four points of interest:

- The resources to be secured are specified as a `web-resource-collection`. In this case, the collection includes any resource available through the path /tcauth, which is the path to TempConvert service deployed in a WAR file. The security thus covers the service's two encapsulated operations, f2c and c2f. This path likewise includes the WSDL, as the URL for the WSDL ends with the path /tcauth?wsdl.

- Access to resources on the path /tcauth is restricted to authenticated users in the role of bigshot. If Fred is to invoke, say, the f2c method, then Fred must have a valid username/password and be authorized to play the role of bigshot.

- The HTTP authentication method is BASIC rather than one of the other standard HTTP methods: DIGEST, FORM, and CLIENT-CERT. Each of these will be clarified shortly. The term *authorization* is used here in the broad sense to cover both user authentication and role authorization.

- The transport is guaranteed to be CONFIDENTIAL, which covers the standard HTTPS services of peer authentication, data encryption, and message integrity. If a user tried to access the resource through an HTTP-based URL such as *http://localhost:8080/tc/tcauth*, Tomcat would then redirect this request to the HTTPS-based URL *https://localhost:8443/tc/tcauth*. (The redirect URL is one of the configuration points specified in *conf/server.xml*.)

The *web.xml* configuration allows settings per HTTP verb, if needed. For example, tight security could be set on POST requests, but security could be lifted altogether for GET requests.

The web service implementation is straightforward:

```
package ch05.tcauth;

import javax.jws.WebService;

@WebService(endpointInterface = "ch05.tcauth.TempConvertAuth")
```

```
public class TempConvertAuthImpl implements TempConvertAuth {
    public float c2f(float t) { return 32.0f + (t * 9.0f / 5.0f); }
    public float f2c(float t) { return (5.0f / 9.0f) * (t - 32.0f); }
}
```

Equally straightforward is this sample client against the service:

```
import tcauthClient.TempConvertAuthImplService;
import tcauthClient.TempConvertAuth;
import javax.xml.ws.BindingProvider;

class ClientAuth {
    public static void main(String args[ ]) {
        TempConvertAuthImplService service = new TempConvertAuthImplService();
        TempConvertAuth port = service.getTempConvertAuthImplPort();
        BindingProvider prov = (BindingProvider) port;

        prov.getRequestContext().put(BindingProvider.USERNAME_PROPERTY, "fred");
        prov.getRequestContext().put(BindingProvider.PASSWORD_PROPERTY, "rockbed");

        System.out.println("f2c(-40.1) ==> " + port.f2C(-40.1f));
        System.out.println("c2f(-40.1) ==> " + port.c2F(-40.1f));
        System.out.println("f2c(+98.7) ==> " + port.f2C(+98.7f));
    }
}
```

There is one tricky aspect to the client: the use of the *wsimport* utility to generate the artifacts in the tcauthClient package. The problem is that the web service's WSDL is also secured and therefore requires authentication for access. There are workarounds, of course. One option is to generate the WSDL locally by using the *wsgen* utility on the SIB. Another option is to get the WSDL from a nonsecure version of the service. The locally saved WSDL and its XSD are then fed into *wsimport* to generate the artifacts.

The client application ClientAuth uses the BindingProvider constants as keys for the username and password. Tomcat expects the lookup keys for the username and password to be the strings:

```
javax.xml.ws.security.auth.username
javax.xml.ws.security.auth.password
```

These are the values of the BindingProvider constant USERNAME_PROPERTY and the constant PASSWORD_PROPERTY, respectively.

When Tomcat receives the request for the secured resource, Tomcat knows from the WAR file's configuration document, *web.xml*, that the requester needs to be authenticated and authorized. Tomcat then checks the submitted username and the associated password credential against data stored in the default MemoryRealm, which contains the usernames, passwords, and authorization roles from the file *conf/tomcat-users.xml*. Here is the file for this example:

```
<?xml version='1.0' encoding='utf-8'?>
<tomcat-users>
  <role rolename="tomcat"/>
  <role rolename="bigshot"/>
```

```
<user username="tomcat" password="tomcat" roles="tomcat"/>
<user username="fred" password="rockbed" roles="bigshot""/>
</tomcat-users>
```

Using a Digested Password Instead of a Password

In addition to BASIC authentication, HTTP 1.1 also supports DIGEST, FORM, and CLIENT-CERT authentication. The FORM method of authentication is the best practice in browser-based web applications. The gist is that the web application itself rather than the browser provides the input form for a username and a password. The CLIENT-CERT method, of course, authenticates the client with a digital certificate but encounters serious practical problems. Imagine, for example, that a client has a good certificate on one machine but unexpectedly must access the secured website from a different machine, which does not have a copy of the certificate.

Tomcat also supports the DIGEST method of authentication. A careful user will be understandably concerned about having a copy of a password on a remote host, in this case the machine that hosts the secured web service. If the remote host is compromised, so is the user's password. A digest of the password avoids this problem so long as the digest is generated with a *one-way* hash function; that is, a function that is easy to compute but hard to invert. For example, given the digest of a password and the algorithm used to compute the digest (for instance, MD5 or SHA-1), it is still a computationally intractable task to derive the original password from the digest produced by a one-way hash function used in MD5 or SHA-1.

Switching from BASIC to DIGEST authentication is not hard. Technically, the *web.xml* entry:

```
<login-config>
    <auth-method>BASIC</auth-method>
</login-config>
```

should be changed to:

```
<login-config>
    <auth-method>DIGEST</auth-method>
</login-config>
```

and a comparable change should be made in the *conf/tomcat-users.xml* file for a web service secured with the MemoryRealm, as in this example. However, there is a simpler approach that can be summarized as follows:

- Tomcat provides a *digest* utility, in *TOMCAT_HOME/bin*, that can be used to generate message digests using the standard algorithms such as MD5 and SHA. Here is the command to digest Fred's password rockbed:

    ```
    % digest.sh -a SHA rockbed
    ```

    There is a *digest.bat* for Windows. The output is a 20-byte digest, in this case:

    ```
    4b177c8995e6b0fa796581ac191f256545f0b8c5
    ```

- The digested password now replaces the password in *tomcat-users.xml*, which becomes:

```
<?xml version='1.0' encoding='utf-8'?>
<tomcat-users>
  <role rolename="tomcat"/>
  <role rolename="bigshot"/>
  <user username="tomcat" password="tomcat" roles="tomcat"/>
  <user username="fred"
        password="4b177c8995e6b0fa796581ac191f256545f0b8c5" roles="bigshot"/>
</tomcat-users>
```

- A client such as Fred now needs to generate the same digest. Tomcat provides a `RealmBase.Digest` method, which the Tomcat *digest* utility uses. Here is the revised client:

```
import tcauthClient.TempConvertAuthImplService;
import tcauthClient.TempConvertAuth;
import javax.xml.ws.BindingProvider;
import org.apache.catalina.realm.RealmBase;

// Revised to send a digested password.
class ClientAuth {
    public static void main(String args[ ]) {
        TempConvertAuthImplService service = new TempConvertAuthImplService();
        TempConvertAuth port = service.getTempConvertAuthImplPort();
        BindingProvider prov = (BindingProvider) port;

        String digest = RealmBase.Digest("rockbed", // password
                                         "SHA",     // digest algorithm
                                         null);     // default char. encoding

        prov.getRequestContext().put(BindingProvider.USERNAME_PROPERTY, "fred");
        prov.getRequestContext().put(BindingProvider.PASSWORD_PROPERTY, digest);

        System.out.println("f2c(-40.1) ==> " + port.f2C(-40.1f));
        System.out.println("c2f(-40.1) ==> " + port.c2F(-40.1f));
        System.out.println("f2c(+98.7) ==> " + port.f2C(+98.7f));
    }
}
```

The revised client needs to be executed with the JAR files *bin/tomcat-juli.jar* and *lib/catalina.jar* on the classpath. The compiler needs only the *catalina.jar* file on the classpath.

Sending a digested password in place of the real thing requires only a little extra work.

A Secured @WebServiceProvider

Tomcat can deploy both a `@WebService` and a `@WebServiceProvider`. Here is the revised `TempConvert` but now as a RESTful service:

```
package ch05.authrest;

import javax.xml.ws.Provider;
import javax.xml.transform.Source;
import javax.xml.transform.stream.StreamSource;
import javax.annotation.Resource;
import javax.xml.ws.BindingType;
import javax.xml.ws.WebServiceContext;
import javax.xml.ws.handler.MessageContext;
import javax.xml.ws.http.HTTPException;
import javax.xml.ws.WebServiceProvider;
import javax.xml.ws.http.HTTPBinding;
import java.io.ByteArrayInputStream;
import java.io.ByteArrayOutputStream;
import java.beans.XMLEncoder;
import java.util.List;
import java.util.ArrayList;

@WebServiceProvider
@BindingType(value = HTTPBinding.HTTP_BINDING)
public class TempConvertR implements Provider<Source> {
    @Resource
    protected WebServiceContext ws_ctx;

    public Source invoke(Source request) {
        // Grab the message context and extract the request verb.
        MessageContext msg_ctx = ws_ctx.getMessageContext();
        String http_verb = (String)
            msg_ctx.get(MessageContext.HTTP_REQUEST_METHOD);
        http_verb = http_verb.trim().toUpperCase();

        if (http_verb.equals("GET")) return doGet(msg_ctx);
        else throw new HTTPException(405); // bad verb exception
    }
    private Source doGet(MessageContext msg_ctx) {
        String query_string = (String) msg_ctx.get(MessageContext.QUERY_STRING);
        if (query_string == null) throw new HTTPException(400); // bad request

        String temp = get_value_from_qs("temp", query_string);
        if (temp == null) throw new HTTPException(400); // bad request

        List<String> converts = new ArrayList<String>();
        try {
            float f = Float.parseFloat(temp.trim());
            float f2c = f2c(f);
            float c2f = c2f(f);
            converts.add(f2c + "C");
            converts.add(c2f + "F");
        }
        catch (NumberFormatException e) { throw new HTTPException(400); }

        // Generate XML and return.
        ByteArrayInputStream stream = encode_to_stream(converts);
        return new StreamSource(stream);
    }
```

```
    private String get_value_from_qs(String key, String qs) {
        String[ ] parts = qs.split("=");
        if (!parts[0].equalsIgnoreCase(key))
            throw new HTTPException(400); // bad request
        return parts[1].trim();
    }

    private ByteArrayInputStream encode_to_stream(Object obj) {
        // Serialize object to XML and return
        ByteArrayOutputStream stream = new ByteArrayOutputStream();
        XMLEncoder enc = new XMLEncoder(stream);
        enc.writeObject(obj);
        enc.close();
        return new ByteArrayInputStream(stream.toByteArray());
    }
    private float c2f(float t) { return 32.0f + (t * 9.0f / 5.0f); }
    private float f2c(float t) { return (5.0f / 9.0f) * (t - 32.0f); }
}
```

Nothing in the code indicates that authentication and authorization are in play, as these security tasks again have delegated to the Tomcat container. The code implements only application logic. The *web.xml* and *sun-jaxws.xml* deployment files are are not changed in any important way.

The revised client against the @WebServiceProvider takes the same approach as the original client against the @WebService—the username and digested password are inserted into the request context with the keys that Tomcat expects:

```
import javax.xml.namespace.QName;
import javax.xml.ws.Service;
import javax.xml.ws.Dispatch;
import javax.xml.ws.http.HTTPBinding;
import org.xml.sax.InputSource;
import javax.xml.xpath.XPath;
import javax.xml.xpath.XPathFactory;
import javax.xml.transform.Source;
import javax.xml.transform.stream.StreamSource;
import javax.xml.ws.handler.MessageContext;
import javax.xml.ws.BindingProvider;
import org.apache.catalina.realm.RealmBase;

class DispatchClientTC {
    public static void main(String[ ] args) throws Exception {
        QName service_name = new QName("TempConvert");
        QName port_name = new QName("TempConvertPort");
        String endpoint = "https://localhost:8443/tempcR/authRest";

        // Now create a service proxy or dispatcher.
        Service service = Service.create(service_name);
        service.addPort(port_name, HTTPBinding.HTTP_BINDING, endpoint);
        Dispatch<Source> dispatch =
            service.createDispatch(port_name, Source.class, Service.Mode.PAYLOAD);
```

```
String digest = RealmBase.Digest("rockbed", // password
                                  "SHA",     // digest algorithm
                                  null);     // default encoding

dispatch.getRequestContext().put(BindingProvider.USERNAME_PROPERTY,
                                 "fred");
dispatch.getRequestContext().put(BindingProvider.PASSWORD_PROPERTY,
                                 digest);
dispatch.getRequestContext().put(MessageContext.HTTP_REQUEST_METHOD, "GET");
dispatch.getRequestContext().put(MessageContext.QUERY_STRING, "temp=-40.1");

StreamSource result = (StreamSource) dispatch.invoke(null);
InputSource source = new InputSource(result.getInputStream());
String expression = "//object";
XPath xpath = XPathFactory.newInstance().newXPath();
String list = xpath.evaluate(expression, source);
System.out.println(list);
    }
}
```

The output is:

```
-40.055557C
-40.18F
```

WS-Security

WS-Security is a family of specifications (see Figure 5-6) designed to augment wire-level security by providing a unified, transport-neutral, end-to-end framework for higher levels of security such as authentication and authorization.

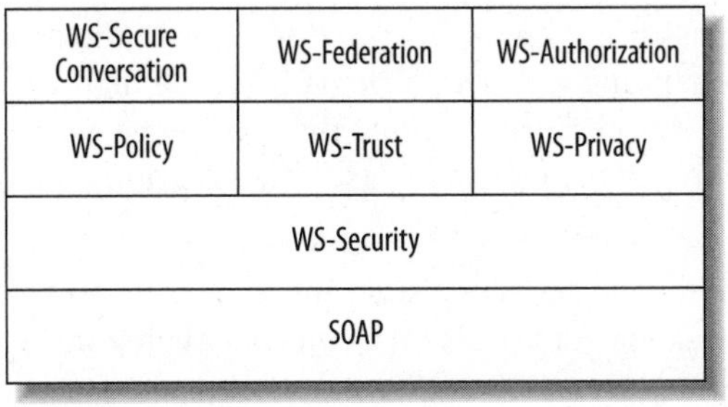

WS-Secure Conversation	WS-Federation	WS-Authorization
WS-Policy	WS-Trust	WS-Privacy
WS-Security		
SOAP		

Figure 5-6. The WS-Security specifications

The layered blocks above WS-Security in Figure 5-6 can be clarified briefly as follows. The first layer consists of WS-Policy, WS-Trust, and WS-Privacy. The second layer of WS-SecureConversation, WS-Federation, and WS-Authorization builds upon this first layer. The architecture is thus modular but also complicated. Here is a short description of each specification, starting with the first layer:

WS-Policy

This specification describes general security capabilities, constraints, and policies. For example, a WS-Policy assertion could stipulate that a message requires security tokens or that a particular encryption algorithm be used.

WS-Trust

This specification deals primarily with how security tokens are to be issued, renewed, and validated. In general, the specification covers *broker trust relationships*, which are illustrated later in a code example.

WS-Privacy

This specification explains how services can state and enforce privacy policies. The specification also covers how a service can determine whether a requester intends to follow such policies.

WS-SecureConversation

This specification covers, as the name indicates, secure web service conversations across different sites and, therefore, across different security contexts and trust domains. The specification focuses on how a security context is created and how security keys are derived and exchanged.

WS-Federation

This specification addresses the challenge of managing security identities across different platforms and organizations. At the heart of the challenge is how to maintain a single, authenticated identity (for example, Alice's security identity) in a heterogeneous security environment.

WS-Authorization

This specification covers the management of authorization data such as security tokens and underlying policies for granting access to secured resources.

WS-Security is often associated with *federated security* in the broad sense, which has the goal of cleanly separating web service logic from the high-level security concerns, in particular authentication/authorization, that challenge web service deployment. This separation of concerns is meant to ease collaboration across computer systems and trust realms.

Recall that SOAP-based web services are meant to be transport-neutral. Accordingly, SOAP-based services cannot depend simply on the reliable transport that HTTP and HTTPS provide, although most SOAP messages are transported over HTTP. HTTP and HTTPS rest on *TCP/IP* (Transmission Control Protocol/Internet Protocol), which supports reliable messaging. What if TCP/IP infrastructure is not available? The WS-ReliableMessaging specification addresses precisely the issue of delivering SOAP-based services over unreliable infrastructure.

A SOAP-based service cannot rely on the authentication/authorization support that a web container such as Tomcat or an application server such as BEA WebLogic, JBoss, GlassFish, or WebSphere may provide. The WS-Security specifications therefore address issues of high-level security as part of SOAP itself rather than as the part of the infrastructure that happens to be in place for a particular SOAP-based service. The goals of WS-Security are often summarized with the phrase *end-to-end* security, which means that security matters are not delegated to the transport level but rather handled directly through an appropriate security API. A framework for end-to-end security needs to cover the situation in which a message is routed through intermediaries, each of which may have to process the message, before reaching the ultimate receiver. Accordingly, end-to-end security focuses on message content rather than on the underlying transport.

Securing a @WebService with WS-Security Under Endpoint

Herein is the source code for a barebones service that will be secured with WS-Security:

```
package ch05.wss;

import javax.jws.WebService;
import javax.jws.WebMethod;
import javax.xml.ws.WebServiceContext;
import javax.annotation.Resource;

@WebService
public class Echo {
    @Resource
    WebServiceContext ws_ctx;

    @WebMethod
    public String echo(String msg) {
        return "Echoing: " + msg;
    }
}
```

Nothing in the code hints at WSS security. The publisher code provides the first hint:

```
package ch05.wss;

import javax.xml.ws.Endpoint;
import javax.xml.ws.Binding;
import javax.xml.ws.soap.SOAPBinding;
import java.util.List;
import java.util.LinkedList;
import javax.xml.ws.handler.Handler;

public class EchoPublisher {
    public static void main(String[ ] args) {
        Endpoint endpoint = Endpoint.create(new Echo());
        Binding binding = endpoint.getBinding();
        List<Handler> hchain = new LinkedList<Handler>();
        hchain.add(new EchoSecurityHandler());
        binding.setHandlerChain(hchain);
        endpoint.publish("http://localhost:7777/echo");
        System.out.println("http://localhost:7777/echo");
    }
}
```

Note that there is a programmatically added handler. Here is the code:

```
package ch05.wss;

import java.util.Set;
import java.util.HashSet;
import javax.xml.namespace.QName;
import javax.xml.soap.SOAPMessage;
import javax.xml.ws.handler.MessageContext;
import javax.xml.ws.handler.soap.SOAPHandler;
import javax.xml.ws.handler.soap.SOAPMessageContext;
import java.io.FileInputStream;
import java.io.File;
import com.sun.xml.wss.ProcessingContext;
import com.sun.xml.wss.SubjectAccessor;
import com.sun.xml.wss.XWSSProcessorFactory;
import com.sun.xml.wss.XWSSProcessor;
import com.sun.xml.wss.XWSSecurityException;

public class EchoSecurityHandler implements SOAPHandler<SOAPMessageContext> {
    private XWSSProcessor xwss_processor = null;
    private boolean trace_p;

    public EchoSecurityHandler() {
        XWSSProcessorFactory fact = null;
        try {
            fact = XWSSProcessorFactory.newInstance();
        }
        catch(XWSSecurityException e) { throw new RuntimeException(e); }

        FileInputStream config = null;
        try {
            config =  new FileInputStream(new File("META-INF/server.xml"));
            xwss_processor =
                fact.createProcessorForSecurityConfiguration(config, new Verifier());
            config.close();
        }
        catch (Exception e) { throw new RuntimeException(e); }
        trace_p = true; // set to true to enable message dumps
    }

    public Set<QName> getHeaders() {
        String uri = "http://docs.oasis-open.org/wss/2004/01/" +
                     "oasis-200401-wss-wssecurity-secext-1.0.xsd";
        QName security_hdr = new QName(uri, "Security", "wsse");
        HashSet<QName> headers = new HashSet<QName>();
        headers.add(security_hdr);
        return headers;
    }

    public boolean handleMessage(SOAPMessageContext msg_ctx) {
        Boolean outbound_p = (Boolean)
            msg_ctx.get (MessageContext.MESSAGE_OUTBOUND_PROPERTY);
        SOAPMessage msg = msg_ctx.getMessage();

        if (!outbound_p.booleanValue()) {
```

```
        // Validate the message.
        try {
          ProcessingContext p_ctx =  xwss_processor.createProcessingContext(msg);
          p_ctx.setSOAPMessage(msg);
          SOAPMessage verified_msg = xwss_processor.verifyInboundMessage(p_ctx);
          msg_ctx.setMessage(verified_msg);

          System.out.println(SubjectAccessor.getRequesterSubject(p_ctx));
          if (trace_p) dump_msg("Incoming message:", verified_msg);
        }
        catch (XWSSecurityException e) { throw new RuntimeException(e); }
        catch(Exception e) { throw new RuntimeException(e); }
      }
      return true;
  }

  public boolean handleFault(SOAPMessageContext msg_ctx) { return true; }
  public void close(MessageContext msg_ctx) { }

  private void dump_msg(String msg, SOAPMessage soap_msg) {
      try {
          System.out.println(msg);
          soap_msg.writeTo(System.out);
          System.out.println();
      }
      catch(Exception e) { throw new RuntimeException(e); }
  }
}
```

Two sections of the EchoSecurityHandler are of special interest. The first callback is the getHeaders, which the runtime invokes *before* invoking the handleMessage callback. The getHeaders method generates a security header block that complies with *OASIS* (Organization for the Advancement of Structured Information Standards) standards, in particular the standards for *WSS* (Web Services Security). The security processor validates the security header.

The second section of interest is, of course, the handleMessage callback that does most of the work. The incoming SOAP message (that is, the client's request) is verified by authenticating the client with a username/password check. The details will follow shortly. If the verification succeeds, the verified SOAP message becomes the new request message. If the verification fails, a XWSSecurityException is thrown as a SOAP fault. The code segment is:

```
try {
    ProcessingContext p_ctx =  xwss_processor.createProcessingContext(msg);
    p_ctx.setSOAPMessage(msg);
    SOAPMessage verified_msg = xwss_processor.verifyInboundMessage(p_ctx);
    msg_ctx.setMessage(verified_msg);
```

```
        System.out.println(SubjectAccessor.getRequesterSubject(p_ctx));
        if (trace_p) dump_msg("Incoming message:", verified_msg);
    }
    catch (XWSSecurityException e) { throw new RuntimeException(e); }
```

On a successful verification, the print statement outputs:

```
Subject:
    Principal: CN=fred
    Public Credential: fred
```

where Fred is the authenticated subject with a *principal*, which is a specific identity under users/roles security. (The *CN* stands for Common Name.) Fred's name acts as the public credential, but his password remains secret.

Publishing a WS-Security Service with Endpoint

Web services that use WS-Security require packages that currently do not ship with core Java 6. It is easier to develop and configure such services with an IDE such as NetBeans and to deploy the services with an application server such as GlassFish, an approach taken in Chapter 6.

It is possible to publish a WSS-based service with the Endpoint publisher, as this section illustrates. Here are the setup steps:

- The JAR file *xws-security-3.0.jar* should be downloaded, as the packages therein are not currently part of core Java 6. A convenient site for downloading is *http://fisheye5.cenqua.com/browse/xwss/repo/com.sun.xml.wss*. For convenience, this JAR file can be placed in *METRO_HOME/lib*.

- For compilation and execution, two JAR files should be on the classpath: *jaxws-tools.jar* and *xws-security-3.0.jar*.

- The configuration files for a WSS-based service usually are housed in a *META-INF* subdirectory of the working directory; that is, the directory in which the service's publisher and the client are invoked. In this case, the working directory is the parent directory of the *ch05* subdirectory. There are two configuration files used in this example: *server.xml* and *client.xml*. The two files are identical to keep the example as simple as possible; in production, of course, they likely would differ.

Once the web service is deployed, *wsimport* can be used to generate the usual client-side artifacts. The Metro runtime kicks in automatically to generate the *wsgen* artifacts; hence, *wsgen* need not be run manually.

The example illustrates the clean separation of security concerns and application logic. All of the WS-Security code is confined to handlers on the client side and the service side.

The EchoSecurityHandler has a no-argument constructor that creates a security processor from information in the configuration file, in this case *server.xml*. Here is the constructor:

```
public EchoSecurityHandler() {
    XWSSProcessorFactory fact = null;
    try {
        fact = XWSSProcessorFactory.newInstance();
    }
    catch(XWSSecurityException e) { throw new RuntimeException(e); }

    FileInputStream config = null;
    try {
        config =  new FileInputStream(new File("META-INF/server.xml"));
        xwss_processor =
        fact.createProcessorForSecurityConfiguration(config, new Verifier());
        config.close();
    }
    catch (Exception e) { throw new RuntimeException(e); }
    trace_p = true; // set to true to enable message dumps
}
```

The Verifier object in the highlighted line does the actual validation. The configuration file is very simple:

```
<!-- Copyright 2004 Sun Microsystems, Inc. All rights reserved.
     SUN PROPRIETARY/CONFIDENTIAL. Use is subject to license terms. -->
<xwss:SecurityConfiguration xmlns:xwss="http://java.sun.com/xml/ns/xwss/config"
                            dumpMessages="true" >
    <xwss:RequireUsernameToken passwordDigestRequired="false"/>
</xwss:SecurityConfiguration>
```

and usually is stored in a *META-INF* directory. In this example, the configuration file stipulates that messages to and from the web service should be dumped for inspection and that the password rather than a password digest is acceptable in the request message. The Verifier object handles the low-level details of authenticating the request by validating the incoming username and password. Here is the code:

```
package ch05.wss;

import javax.security.auth.callback.Callback;
import javax.security.auth.callback.CallbackHandler;
import javax.security.auth.callback.UnsupportedCallbackException;
import com.sun.xml.wss.impl.callback.PasswordCallback;
import com.sun.xml.wss.impl.callback.PasswordValidationCallback;
import com.sun.xml.wss.impl.callback.UsernameCallback;

// Verifier handles service-side callbacks for password validation.
public class Verifier implements CallbackHandler {
    // Username/password hard-coded for simplicity and clarity.
    private static final String _username = "fred";
    private static final String _password = "rockbed";

    // For password validation, set the validator to the inner class below.
    public void handle(Callback[ ] callbacks) throws UnsupportedCallbackException {
        for (int i = 0; i < callbacks.length; i++) {
            if (callbacks[i] instanceof PasswordValidationCallback) {
                PasswordValidationCallback cb=(PasswordValidationCallback)callbacks[i];
                if (cb.getRequest() instanceof
```

```
                    PasswordValidationCallback.PlainTextPasswordRequest)
                    cb.setValidator(new PlainTextPasswordVerifier());
            }
            else
                throw new UnsupportedCallbackException(null, "Not needed");
        }
    }

    // Encapsulated validate method verifies the username/password.
    private class PlainTextPasswordVerifier
                    implements PasswordValidationCallback.PasswordValidator {
        public boolean validate(PasswordValidationCallback.Request req)
            throws PasswordValidationCallback.PasswordValidationException {

            PasswordValidationCallback.PlainTextPasswordRequest plain_pwd =
                (PasswordValidationCallback.PlainTextPasswordRequest) req;
            if (_username.equals(plain_pwd.getUsername()) &&
                _password.equals(plain_pwd.getPassword())) {
                return true;
            }
            else
                return false;
        }
    }
}
```

The authentication succeeds with a username of fred and a password of rockbed but
fails otherwise. The security processor created in the EchoSecurityHandler constructor
is responsible for invoking the handle callback in the Verifier. To keep the example
simple, the username and password are hardcoded rather than retrieved from a database. Further, the password is plain text.

A client-side SOAPHandler generates and inserts the WSS artifacts that the Echo service
expects. Here is the source code:

```
    package ch05.wss;

    import java.util.Set;
    import java.util.HashSet;
    import javax.xml.namespace.QName;
    import javax.xml.soap.SOAPMessage;
    import javax.xml.ws.handler.MessageContext;
    import javax.xml.ws.handler.soap.SOAPHandler;
    import javax.xml.ws.handler.soap.SOAPMessageContext;
    import java.io.FileInputStream;
    import java.io.File;
    import com.sun.xml.wss.ProcessingContext;
    import com.sun.xml.wss.SubjectAccessor;
    import com.sun.xml.wss.XWSSProcessorFactory;
    import com.sun.xml.wss.XWSSProcessor;
    import com.sun.xml.wss.XWSSecurityException;

    public class ClientHandler implements SOAPHandler<SOAPMessageContext> {
        private XWSSProcessor xwss_processor;
        private boolean trace_p;
```

```
public ClientHandler() {
    XWSSProcessorFactory fact = null;
    try {
        fact = XWSSProcessorFactory.newInstance();
    }
    catch(XWSSecurityException e) { throw new RuntimeException(e); }

    // Read client configuration file and configure security.
    try {
        FileInputStream config =
            new FileInputStream(new File("META-INF/client.xml"));
        xwss_processor =
            fact.createProcessorForSecurityConfiguration(config, new Prompter());
        config.close();
    }
    catch (Exception e) { throw new RuntimeException(e); }
    trace_p = true; // set to true to enable message dumps
}

// Add a security header block
public Set<QName> getHeaders() {
    String uri = "http://docs.oasis-open.org/wss/2004/01/" +
        "oasis-200401-wss-wssecurity-secext-1.0.xsd";
    QName security_hdr = new QName(uri, "Security", "wsse");
    HashSet<QName> headers = new HashSet<QName>();
    headers.add(security_hdr);
    return headers;
}

public boolean handleMessage(SOAPMessageContext msg_ctx) {
    Boolean outbound_p = (Boolean)
        msg_ctx.get (MessageContext.MESSAGE_OUTBOUND_PROPERTY);
    SOAPMessage msg = msg_ctx.getMessage();

    if (outbound_p.booleanValue()) {
        // Create a message that can be validated.
        ProcessingContext p_ctx = null;
        try {
            p_ctx = xwss_processor.createProcessingContext(msg);
            p_ctx.setSOAPMessage(msg);
            SOAPMessage secure_msg = xwss_processor.secureOutboundMessage(p_ctx);
            msg_ctx.setMessage(secure_msg);

            if (trace_p) dump_msg("Outgoing message:", secure_msg);
        }
        catch (XWSSecurityException e) { throw new RuntimeException(e); }
    }
    return true;
}

public boolean handleFault(SOAPMessageContext msg_ctx) { return true; }
public void close(MessageContext msg_ctx) { }
```

```
        private void dump_msg(String msg, SOAPMessage soap_msg) {
            try {
                System.out.println(msg);
                soap_msg.writeTo(System.out);
                System.out.println();
            }
            catch(Exception e) { throw new RuntimeException(e); }
        }
}
```

The Prompter and the Verifier

On the service side, the security processor uses a `Verifier` object to authenticate the request. On the client side, the security processor uses a `Prompter` object to get the username and password, which then are inserted into the outgoing SOAP message. Just as the service-side security processor generates a validated SOAP message if the authentication succeeds, so the client-side security processor generates a secured SOAP message if the `Prompter` works correctly. Here is the `Prompter` code:

```
package ch05.wss;

import java.io.IOException;
import javax.security.auth.callback.Callback;
import javax.security.auth.callback.CallbackHandler;
import javax.security.auth.callback.UnsupportedCallbackException;
import com.sun.xml.wss.impl.callback.PasswordCallback;
import com.sun.xml.wss.impl.callback.PasswordValidationCallback;
import com.sun.xml.wss.impl.callback.UsernameCallback;
import java.io.BufferedReader;
import java.io.InputStreamReader;

// Prompter handles client-side callbacks, in this case
// to prompt for and read username/password.
public class Prompter implements CallbackHandler {
    // Read username or password from standard input.
    private String readLine() {
        String line = null;
        try {
            line = new BufferedReader(new InputStreamReader(System.in)).readLine();

        }
        catch(IOException e) { }
        return line;
    }

    // Prompt for and read the username and the password.
    public void handle(Callback[ ] callbacks) throws UnsupportedCallbackException {
        for (int i = 0; i < callbacks.length; i++) {
            if (callbacks[i] instanceof UsernameCallback) {
                UsernameCallback cb = (UsernameCallback) callbacks[i];
                System.out.print("Username: ");
                String username = readLine();
```

```
            if (username != null) cb.setUsername(username);
        }
        else if (callbacks[i] instanceof PasswordCallback) {
            PasswordCallback cb = (PasswordCallback) callbacks[i];
            System.out.print("Password: ");
            String password = readLine();
            if (password != null) cb.setPassword(password);
        }
      }
    }
}
```

The security processor interacts with the `Prompter` through a `UsernameCallback` and a `PasswordCallback`, which prompt for, read, and store the client's username and password.

The Secured SOAP Envelope

The client-side security processor generates a SOAP message with all of the WSS information in the SOAP header. The SOAP body is indistinguishable from one in an unsecured SOAP message. Here is the SOAP message that the client sends:

```
<?xml version="1.0" encoding="UTF-8"?>
<S:Envelope xmlns:S="http://schemas.xmlsoap.org/soap/envelope/">
<S:Header>
    <wsse:Security xmlns:wsse="http://docs.oasis-open.org/wss/2004/01/
                              oasis-200401-wss-wssecurity-secext-1.0.xsd"
                   S:mustUnderstand="1">
    <wsse:UsernameToken xmlns:wsu="http://docs.oasis-open.org/wss/2004/01/
                                   oasis-200401-wss-wssecurity-utility-1.0.xsd"
                        wsu:Id="XWSSGID-12168512095289497835553">
      <wsse:Username>fred</wsse:Username>
      <wsse:Password
         Type="http://docs.oasis-open.org/wss/2004/01/
               oasis-200401-wss-username-token-profile-1.0#PasswordText">
          rockbed
      </wsse:Password>
      <wsse:Nonce EncodingType="http://docs.oasis-open.org/wss/2004/01/
                               oasis-200401-wss-soap-message-security-1.0#Base64Binary">
          Vyg9oXUn/rl2F4m6lSFIZCoU
      </wsse:Nonce>
      <wsu:Created>2008-07-23T22:13:33.001Z</wsu:Created>
    </wsse:UsernameToken>
    </wsse:Security>
</S:Header>
<S:Body>
    <ns2:echo xmlns:ns2="http://wss.ch05/">
        <arg0>Hello, world!</arg0>
    </ns2:echo>
</S:Body>
</S:Envelope>
```

In addition to the username and the password, the SOAP header includes a *nonce*, which is a randomly generated, statistically unique cryptographic token inserted into the message. A nonce is added to safeguard against message theft and *replay attacks* in which an unsecured credential such as a password is retransmitted to perform an unauthorized operation. For example, if Eve were to intercept Alice's password from a SOAP message that transfers funds from one of Alice's bank accounts to another of Alice's accounts, then Eve might replay the scenario at a later time using the pirated password to transfer Alice's funds into Eve's account. If the message receiver requires not only the usual credentials such as a password but also a nonce with certain secret attributes, then Eve would need more than just the pirated password—Eve also would need to replicate the nonce, which should be computationally intractable.

The SOAP message from the `Echo` service to the client has no WSS artifacts at all:

```
<S:Envelope xmlns:S="http://schemas.xmlsoap.org/soap/envelope/">
  <S:Header/>
  <S:Body>
    <ns2:echoResponse xmlns:ns2="http://wss.ch05/">
      <return>Echoing: Hello, world!</return>
    </ns2:echoResponse>
  </S:Body>
</S:Envelope>
```

Summary of the WS-Security Example

This first example of WS-Security introduces the API but has the obvious drawback of sending the client's username and password, together with the nonce, over an unsecure channel. The reason, of course, is that the `Endpoint` publisher does not support HTTPS connections. A quick fix would be to use the `HttpsServer` illustrated earlier.

Yet the example does meet the goal of illustrating how WS-Security itself, without any support from a transport protocol such as HTTP or a container such as Tomcat, supports authentication and authorization. The example likewise shows that WS-Security encourages a clean separation of security concerns from web service logic. Chapter 6 drills deeper into the details of WS-Security and provides a production-grade example.

What's Next?

So far all of the sample web services have been deployed using either `Endpoint`, `HttpsServer`, or Tomcat. This low-fuss approach has the benefit of keeping the focus on Java's web service API. In a production environment, a lightweight web container such as Tomcat or a heavyweight application server such as BEA WebLogic, GlassFish, JBoss, or WebSphere would be the likely choice for deploying SOAP-based and REST-style services. The next chapter looks at GlassFish, which includes Metro and is the reference implementation of a Java web services container.

## CHAPTER 6

JAX-WS in Java Application Servers

Overview of a Java Application Server

In previous chapters, SOAP-based and REST-style web services have been deployed using mostly the `Endpoint` publisher or the Tomcat web container. This chapter illustrates how web services can be deployed using a Java Application Server (JAS), the software centerpiece of enterprise Java. The current version of enterprise Java is Java EE 5, which includes EJB 3.0. To begin, here is a sketch of the software bundled into a JAS:

Web container

A web container deploys servlets and web services. A traditional web application in Java is a mix of static HTML pages, servlets, higher-level servlet generators such as *JSP* (Java Server Pages) and *JSF* (Java Server Faces) scripts, backend Java-Beans for JSP and JSF scripts, and utility classes. Tomcat is the reference implementation (RI) for a web container. Tomcat, like other web containers, can be embedded in an application server. Web components are deployed in the web container as WAR files, which typically contain the standard configuration document *web.xml* and may contain vendor-specific configuration documents as well (e.g., *sun-jaxws.xml*). To host web services, a web container relies on a servlet interceptor (in the case of Tomcat, a `WSServlet` instance) that mediates between the client and the web service SIB.

Message-oriented middleware

The message-oriented middleware supports *JMS* (Java Message Service), which provides the store-and-forward technologies lumped together under the term *messaging*. JMS supports synchronous and asynchronous messaging styles and two types of message repository: topics, which are akin to bulletin boards in that a *read* operation does not automatically remove a posted message, and queues, which are *FIFO* (First In, First Out) lists in which a *read* operation, by default, removes the read item from the queue. In JMS, a *publisher* publishes messages to a topic and a *sender* sends messages to a queue. A *subscriber* to a topic or a *receiver* on a queue receives such messages either synchronously through a blocking *read* operation or asynchronously through the JMS notification mechanism. JMS topics implement the publisher/subscriber model of messaging, whereas JMS queues implement the point-to-point model.

Enterprise Java Bean (EJB) container

The EJB container holds EJB instances, which are of three types: Session, Entity, and Message-Driven. Session and traditional Entity EJBs are built on a Java *RMI* (Remote Method Invocation) foundation, whereas Message-Driven EJBs are built on a JMS foundation.

The Message-Driven EJB is a JMS `MessageListener` implemented as an EJB. A listener receives an event notification whenever a new message arrives at a topic or a queue to which the listener has subscribed.

A Session EJB typically implements an enterprise application's business logic and interacts as needed with other application components, either local (for instance, other EJBs in the same container) or remote (for instance, clients on a different host). As the name suggests, a Session EJB is designed to maintain a client session. A Session EJB is either stateless or stateful. A stateless Session EJB is, in effect, a collection of mutually independent instance methods that should operate only on data passed in as arguments. The EJB container assumes that a stateless Session EJB instance does not maintain state information in instance fields. Suppose, for example, that a Session EJB encapsulates two instance methods, `m1` and `m2`. If this EJB were deployed as stateless, then the EJB container would assume that a particular client, `C`, could invoke `m1` in one EJB instance and `m2` in another EJB instance because the two methods do not share state. If the same Session EJB were deployed as stateful, then the EJB container would have to ensure that `C`'s invocation of `m1` and `m2` involved the *same* EJB instance because the two methods presumably share state. As this summary implies, an EJB container automatically manages a pool of EJB instances for all types of EJB. A SOAP-based web service can be implemented as a *stateless* Session EJB by annotating the SIB with `@Stateless`.

Prior to Java EE 5, an Entity EJB instance was the preferred way to provide an enterprise application with an in-memory cache of database objects such as a table row. The Entity EJB was the persistence construct that brought *ORM* (Object Relational Mapping) capabilities to the application. A traditional Entity EJB could be deployed with either *BMP* (Bean Managed Persistence) or *CMP* (Container Managed Persistence). At issue was whether the programmer or the EJB container maintained coherence between the data source (for instance, a table row) and the EJB instance. In the early days of EJB containers, the case could be made that BMP was more efficient. Yet the EJB containers quickly improved to the point that CMP became the obvious choice. Indeed, CMP emerged as a major inducement for and benefit of using traditional Entity EJBs. An EJB deployed with CMP also had to deploy with *CMT* (Container Managed Transactions).

Prior to EJB 3.0, EJBs in general and Entity EJBs in particular were very difficult to code and to configure. The coding was tricky and the configuration required a large, complicated XML document known affectionately as the *DD* (Deployment Descriptor). All of this changed with Java EE 5, which extended the capabilities of the original Entity EJB to POJO classes annotated with `@Entity`. In effect, the `@Entity` annotation let Java programmers enjoy the benefits of the traditional Entity EJB without enduring the pain of configuring and programming this kind of EJB. The `@Entity` annotation is at the center of the Java Persistence API (JPA), which integrates features from related technologies such as Hibernate, Oracle's TopLink, Java Data Objects, and traditional Entity EJBs. The `@Entity` is now the preferred way to handle persistence. An `@Entity`, unlike an Entity EJB, can be used in either core Java or enterprise Java applications.

EJBs, unlike servlets, are thread-safe because the EJB container assumes responsibility for thread synchronization. (As in the case of servlets, each client request against an EJB executes as a separate thread.) Even in a traditional browser-based web application, EJBs are thus well suited as backend support for servlets. For instance, a servlet might pass a request along to a Session EJB, which in turn might use instances of various `@Entity` classes as persisted data sources (see Figure 6-1).

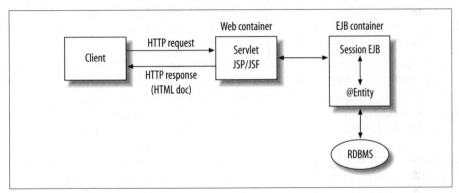

Figure 6-1. Architecture of an enterprise application

JNDI service provider

A *JNDI* (Java Naming and Directory Interface) provider maintains, at the least, a naming service in which names (for instance, the deployed name of an EJB or a message queue) are bound to entities (in this case, the deployed EJB or the queue). If the JNDI provider supports directory services as well, then the name of an entity binds to the entity and its attributes. In the simple case, a JNDI provider maintains a hierarchical database of name/entity pairs. Any component dropped into a JAS container is automatically registered with the naming service and thereafter available for lookup by potential clients. In Java EE 5, the JNDI provider is largely unseen infrastructure.

Security provider

The security provider provides high-level security services for components deployed in any of the containers. Among these services are, of course, authentication and authorization. The security provider has to be *JAAS* (Java Authentication and Authorization Service) compliant. Nowadays a security provider typically has plugins for providers such as an *LDAP* (Lightweight Directory Access Protocol) provider. The security in an application server is usually integrated. For instance, the container-managed security that the web container provides is integrated into the default JAAS-based security that the EJB container provides.

Relational Database Management System

An application server usually includes an *RDMS* (Relational Database Management System), which is the default persistence store for `@Entity` instances and may be accessed programatically and directly from other components such as web services and servlets. A modern application server makes it easy to plug in the database system of choice if the provided system is not the preferred one. The GlassFish application server introduced in this chapter ships with Java DB, Sun's supported distribution of the Apache Derby relational database management system (RDBMS) (*http://db.apache.org/derby*).

Client container

The client container consists of the software libraries that a client requires to interact with deployed components such as message topics or EJBs and to use services such as JNDI and security.

On the *web tier*, a JAS supports both traditional interactive websites and web services. The model-view-controller (MVC) architecture is popular in modern interactive websites. A *model* maintains state information about the site and is responsible for persistence; a *view* provides an appropriate representation of a model; and a *controller* is a request endpoint that implements the business logic that coordinates a model and a view. In enterprise and even in core Java, an `@Entity` instance is a natural way to implement a model with persistence. In interactive web applications, JSP or JSF scripts can generate an HTML view of a model, and either a servlet or a Session EJB is a natural way to implement a controller. For Java-based web services deployed in an application server such as GlassFish, `@Entity` instances are likewise a natural way to implement models. The web service is the controller that exposes business logic in `@WebMethod` code and interacts, as appropriate, with models. The web service can be deployed in a WAR file, which amounts to a servlet-based deployment, or in an *EAR* (Enterprise ARchive) file, which is an EJB-based deployment. The distinction between servlet-based and EJB-based deployment will be discussed again later.

As a bundle of so many features and services, an application server is inherently complicated software. At issue among Java programmers is whether the benefits that come with an application server offset the complexity of using it. This complexity stems, in large part, from the fact that so many APIs come into play in an application server. An application that incorporates servlets, JSP scripts, JSF scripts, messaging, and EJBs must deal with at least five distinct APIs. Integrating web components such as servlets with EJBs remains nontrivial. This state of affairs accounts for recent efforts among vendors to provide seamless integration of Java EE components, which presumably would result in a lighter-weight, more programmer-friendly framework for doing enterprise Java. JBoss Seam is one such effort. It should be noted, however, that Java EE 5 is significantly easier to use than its predecessor, J2EE 1.4. Java EE is definitely moving down the road that lighter frameworks such as Spring cut out for enterprise Java.

The GlassFish application server is open source and the reference implementation. (To be legalistic, a particular *snapshot* of GlassFish is the RI.) The current production release can be downloaded from *https://glassfish.dev.java.net*. This release is also bundled into the NetBeans IDE download, available at *http://www.netbeans.org*. The examples in this chapter are deployed under GlassFish, and some are developed or refined using NetBeans. No example requires the use of a particular IDE, including NetBeans, however.

GlassFish Basics

Assume that `AS_HOME` points to the GlassFish install directory. In the *bin* subdirectory is a script named *asadmin* that can be used to start GlassFish:

```
% asadmin start-domain domain1
```

For Windows, the script is *asadmin.bat*. When the application server starts, it prints the ports on which it listens for requests. For example, port 8080 or port 8081 may be given as the HTTP port. To confirm that GlassFish has started, open a browser to *http://localhost:8080*, assuming that 8080 is among the listed HTTP ports. The welcome page should appear with links to GlassFish documentation and the like. The command:

```
% asadmin stop-domain domain1
```

stops deployment from `domain1`, whereas the command:

```
% asadmin stop-appserv
```

stops the application server as a whole.

The URL *http://localhost:4848* is for the administrative console, which has various utilities for inspecting and managing deployed components, including web services. The default username for the administrative console is `admin`, and the default password is `adminadmin`. The administrative console is helpful in testing newly deployed web services. For any `@WebMethod` that can be invoked through HTTP, GlassFish generates an interactive web page from which the method can be tested with sample arguments. Request and response SOAP messages are displayed. The WSDL and other XML documents are likewise available for inspection.

Deploying a web service is easy. Suppose that a `@WebService` with an SIB class named Hi has been packaged in a WAR file named *hello.war*. (The packaging details are the same as for Tomcat but will be reviewed in the first example later.) To deploy the application, the WAR file is copied to *AS_HOME/domains/domain1/autodeploy*. If the deployment succeeds, a second file named *hello.war_deployed* appears in a second or so in the *autodeploy* directory. If the deployment fails, then the empty file *hello.war_deployedFailed* appears instead. On a failed deployment, the *domains/domain1/logs/server.log* file contains the details. If the deployment succeeds, the WSDL

---

can be inspected by opening a browser to *http://localhost:8080/hello/HiService?wsdl*. The default name of the service, `HiService`, follows the JWS convention of appending `Service` to the SIB name `Hi`.

The Java DB database system can be started with the command:

```
% asadmin start-database
```

This command starts the Derby RDBMS, which then runs independently of the application server. The database is stopped with the command:

```
% asadmin stop-database
```

Deploying @WebServices and @WebServiceProviders

The SOAP-based `Teams` web service from Chapter 1 has four classes: `Player`, `Team`, `TeamsUtility`, and the SIB `Teams`. For this example, the four files reside in the directory *ch06/team* and all are in the Java package `ch06.team`. For review, here is the original SIB but in the new package:

```
package ch06.team;

import java.util.List;
import javax.jws.WebService;
import javax.jws.WebMethod;

@WebService
public class Teams {
    private TeamsUtility utils;
    public Teams() { utils = new TeamsUtility(); }

    @WebMethod
    public Team getTeam(String name) { return utils.getTeam(name); }
    @WebMethod
    public List<Team> getTeams() { return utils.getTeams(); }
}
```

After compilation, the *.class* files are copied to the directory *ch06/team/WEB-INF/classes/ch06/team* because Tomcat, the web container for GlassFish, expects compiled classes to reside in the *WEB-INF/classes* tree. A WAR file is created with the command:

```
% jar cvf team.war WEB-INF
```

and the WAR file then is copied to *AS_HOME/domains/domain1/autodeploy*. Although the `Teams` service is document-style, there is no need to generate manually, using *wsgen*, the JAX-B artifacts that such a service requires. GlassFish ships with the current Metro release, which automatically generates these artifacts.

The client-side *wsimport* artifacts are generated in the usual way:

```
% wsimport -keep -p clientC http://localhost:8081/team/TeamsService?wsdl
```

Here is a sample client that uses the *wsimport*-generated artifacts:

```
import teamsC.TeamsService;
import teamsC.Teams;
import teamsC.Team;
import teamsC.Player;
import java.util.List;

class TeamsClient {
    public static void main(String[ ] args) {
        TeamsService service = new TeamsService();
        Teams port = service.getTeamsPort();

        List<Team> teams = port.getTeams();
        for (Team team : teams) {
            System.out.println("Team name: " + team.getName() +
                               " (roster count: " + team.getRosterCount() + ")");
            for (Player player : team.getPlayers())
                System.out.println(" Player: " + player.getNickname());
        }
    }
}
```

The output is:

```
Team name: Abbott and Costello (roster count: 2)
  Player: Bud
  Player: Lou
Team name: Marx Brothers (roster count: 3)
  Player: Chico
  Player: Groucho
  Player: Harpo
Team name: Burns and Allen (roster count: 2)
  Player: George
  Player: Gracie
```

In addition to the WSDL, GlassFish automatically generates the deployment artifacts, including a *webservices.xml* document and a *sun-web.xml* document. These can be inspected in the administrative console.

Deploying @WebServiceProviders

Deploying a @WebServiceProvider is slightly more complicated in that the WAR file requires a *web.xml* and *sun-jaxws.xml*. Here is source code for a RESTful service that does temperature conversions and computes Fibonacci numbers:

```
package ch06.rest;

import javax.xml.ws.Provider;
import javax.xml.transform.TransformerFactory;
import javax.xml.transform.Transformer;
import javax.xml.transform.Source;
import javax.xml.transform.Result;
import javax.xml.transform.dom.DOMResult;
import javax.xml.transform.stream.StreamSource;
```

```
import javax.xml.transform.stream.StreamResult;
import javax.annotation.Resource;
import javax.xml.ws.BindingType;
import javax.xml.ws.WebServiceContext;
import javax.xml.ws.handler.MessageContext;
import javax.xml.ws.http.HTTPException;
import javax.xml.ws.WebServiceProvider;
import javax.xml.ws.ServiceMode;
import javax.xml.ws.http.HTTPBinding;
import java.io.ByteArrayInputStream;
import java.io.ByteArrayOutputStream;
import java.io.StringWriter;
import org.w3c.dom.Document;
import org.w3c.dom.NodeList;
import org.w3c.dom.Node;
import javax.xml.parsers.DocumentBuilderFactory;
import javax.xml.parsers.DocumentBuilder;

@WebServiceProvider
@ServiceMode(value = javax.xml.ws.Service.Mode.MESSAGE)
@BindingType(value = HTTPBinding.HTTP_BINDING)
public class RestfulProviderD implements Provider<Source> {
    @Resource
    protected WebServiceContext ws_context;
    protected Document document; // DOM tree
    public Source invoke(Source request) {
        try {
            if (ws_context == null) throw new RuntimeException("No ws_context.");
            MessageContext msg_context = ws_context.getMessageContext();
            // Check the HTTP request verb. In this case, only POST is supported.
            String http_verb = (String)
                msg_context.get(MessageContext.HTTP_REQUEST_METHOD);
            if (!http_verb.toUpperCase().trim().equals("POST"))
                throw new HTTPException(405); // bad verb exception
            build_document(request);
            String operation = extract_node("operation").trim();
            String operand = extract_node("operand").trim();
            if (operation.equals("fib"))      return fib_response(operand);
            else if (operation.equals("c2f")) return c2f_response(operand);
            else if (operation.equals("f2c")) return (f2c_response(operand));
            throw new HTTPException(404); // client error
        }
        catch(Exception e) { throw new HTTPException(500); }
    }
    // Build a DOM tree from the XML source for later lookups.
    private void build_document(Source request) {
        try {
            Transformer transformer = TransformerFactory.newInstance().newTransformer();
            this.document =
                DocumentBuilderFactory.newInstance().newDocumentBuilder().newDocument();
            Result result = new DOMResult(this.document);
            transformer.transform(request, result);
        }
        catch(Exception e) { this.document = null; }
    }
```

```
// Extract a node's value from the DOM tree given the node's tag name.
private String extract_node(String tag_name) {
    try {
        NodeList nodes = this.document.getElementsByTagName(tag_name);
        Node node = nodes.item(0);
        return node.getFirstChild().getNodeValue().trim();
    }
    catch(Exception e) { return null; }
}

// Prepare a response Source in which obj refers to the return value.
private Source prepare_source(Object obj) {
    String xml =
        "<uri:restfulProvider xmlns:uri = 'http://foo.bar.baz'>" +
        "<return>" + obj + "</return>" +
        "</uri:restfulProvider>";
    return new StreamSource(new ByteArrayInputStream(xml.getBytes()));
}

private Source fib_response(String num) {
    try {
        int n = Integer.parseInt(num.trim());
        int fib = 1;
        int prev = 0;

        for (int i = 2; i <= n; ++i) {
            int temp = fib;
            fib += prev;
            prev = temp;
        }
        return prepare_source(fib);
    }
    catch(Exception e) { throw new HTTPException(500); }
}

private Source c2f_response(String num) {
    try {
        float c = Float.parseFloat(num.trim());
        float f = 32.0f + (c * 9.0f / 5.0f);
        return prepare_source(f);
    }
    catch(Exception e) { throw new HTTPException(500); }
}

// Compute f2c(c)
private Source f2c_response(String num) {
    try {
        float f = Float.parseFloat(num.trim());
        float c = (5.0f / 9.0f) * (f - 32.0f);
        return prepare_source(c);
    }
    catch(Exception e) { throw new HTTPException(500); }
}
}
```

If the compiled code in *WEB-INF/classes/ch06/rest* were put into a WAR file and deployed without a *web.xml* document and a *sun-jaxws.xml* document, as in the example just shown, a client would get an HTTP 404 (Not Found) error on a request for the service. The `@WebServiceProvider` must be deployed with the two configuration documents. Following is the *web.xml*:

```
<?xml version="1.0" encoding="UTF-8"?>
<web-app version="2.4" xmlns="http://java.sun.com/xml/ns/j2ee"
         xmlns:xsi="http://www.w3.org/2001/XMLSchema-instance"
         xsi:schemaLocation="http://java.sun.com/xml/ns/j2ee
         http://java.sun.com/xml/ns/j2ee/web-app_2_4.xsd">
  <display-name>RestfulProvider Service</display-name>
  <listener>
    <listener-class>
        com.sun.xml.ws.transport.http.servlet.WSServletContextListener
    </listener-class>
  </listener>
  <servlet>
    <display-name>RestfulProviderD</display-name>
    <servlet-name>RestfulProviderD</servlet-name>
    <servlet-class>com.sun.xml.ws.transport.http.servlet.WSServlet</servlet-class>
  </servlet>
  <servlet-mapping>
    <servlet-name>RestfulProviderD</servlet-name>
    <url-pattern>/restful/*</url-pattern>
  </servlet-mapping>
</web-app>
```

And here is the *sun-jaxws.xml*:

```
<?xml version="1.0" encoding="UTF-8"?>
<endpoints xmlns="http://java.sun.com/xml/ns/jax-ws/ri/runtime" version="2.0">
    <endpoint name="RestfulProviderD"
        implementation="ch06.rest.RestfulProviderD"
        binding="http://www.w3.org/2004/08/wsdl/http"
        url-pattern='/restful/*'/>
</endpoints>
```

These are essentially the same XML configuration files, with respect to structure, that were used to deploy SOAP-based and REST-style services under Tomcat. GlassFish automatically recognizes a `@WebService` even without the configuration files but does not similarly recognize a `@WebServiceProvider`; hence, the configuration documents are needed in the second case.

Here is a sample client against the service. For variety, the client does a DOM parse but without using XPath:

```
import javax.xml.ws.Service;
import javax.xml.namespace.QName;
import javax.xml.ws.http.HTTPBinding;
import javax.xml.ws.ServiceMode;
import javax.xml.ws.Dispatch;
import javax.xml.transform.TransformerFactory;
import javax.xml.transform.Transformer;
```

```
import javax.xml.transform.Source;
import javax.xml.transform.stream.StreamSource;
import javax.xml.transform.stream.StreamResult;
import javax.xml.ws.handler.MessageContext;
import java.net.URL;
import java.util.Map;
import java.io.StringReader;
import java.io.ByteArrayOutputStream;
import java.io.ByteArrayInputStream;
import org.w3c.dom.Document;
import org.w3c.dom.NodeList;
import org.w3c.dom.Node;
import javax.xml.parsers.DocumentBuilderFactory;
import javax.xml.parsers.DocumentBuilder;

class DispatchClient {
    private static String xml =
        "<?xml version = '1.0' encoding = 'UTF-8' ?>" + "\n" +
        "<uri:RequestDocument xmlns:uri = 'urn:RequestDocumentNS'>" + "\n" +
        "  <operation>f2c</operation>" + "\n" +
        "  <operand>-40</operand>" + "\n" +
        "</uri:RequestDocument>" + "\n";

    public static void main(String[ ] args) throws Exception {
        QName qname = new QName("", "");
        String url_string = "http://127.0.0.1:8080/restfulD/restful/";
        URL url = new URL(url_string);

        // Create the service and add a port
        Service service = Service.create(qname);
        service.addPort(qname, HTTPBinding.HTTP_BINDING, url_string);

        Dispatch<Source> dispatcher = service.createDispatch(qname,
                                          Source.class,
                                          javax.xml.ws.Service.Mode.MESSAGE);
        Map<String, Object> rc = dispatcher.getRequestContext();
        rc.put(MessageContext.HTTP_REQUEST_METHOD, "POST");
        Source result = dispatcher.invoke(new StreamSource(new StringReader(xml)));
        parse_response(result);
    }

    private static void parse_response(Source res) throws Exception {
        Transformer transformer = TransformerFactory.newInstance().newTransformer();
        ByteArrayOutputStream bao = new ByteArrayOutputStream();
        StreamResult sr = new StreamResult(bao);
        transformer.transform(res, sr);
        ByteArrayInputStream bai = new ByteArrayInputStream(bao.toByteArray());
        DocumentBuilder db = DocumentBuilderFactory.newInstance().newDocumentBuilder();
        Document root = db.parse(bai);
        NodeList nodes = root.getElementsByTagName("return");
        Node node = nodes.item(0); // should be only one <return> element
        System.out.println("Request document:\n" + xml);
        System.out.println("Return value: " + node.getFirstChild().getNodeValue());
    }
}
```

The output is:

```
Request document:
<?xml version = '1.0' encoding = 'UTF-8' ?>
<uri:RequestDocument xmlns:uri = 'urn:RequestDocumentNS'>
  <operation>f2c</operation>
  <operand>-40</operand>
</uri:RequestDocument>

Return value: -40.0
```

The deployment of a @WebService and that of a @WebServiceProvider differ slightly under GlassFish, although the two deployments are the same under either Endpoint or stand-alone Tomcat. The next section illustrates how a deployed web service can be invoked from a JSP or JSF script.

Integrating an Interactive Website and a Web Service

A Web service client is usually not interactive and, in particular, not a browser. This section illustrates, as proof of concept, how a browser-based client might be used with a SOAP-based web service. The @WebService is simple and familiar:

```java
package ch06.tc;
import javax.jws.WebService;
import javax.jws.WebMethod;


@WebService
public class TempConvert {
    @WebMethod
    public float c2f(float t) { return 32.0F + (t * 9.0F / 5.0F); }
    @WebMethod
    public float f2c(float t) { return (5.0F / 9.0F) * (t - 32.0F); }
}
```

After the compiled TempConvert is copied to the *ch06/tc/WEB-INF/classes/ch06/tc* directory, the WAR file is created and copied as usual to the *domains/domain1/autodeploy* directory under *AS_HOME*.

The next step is to write the interactive web application, which in this case consists of an HTML document, two small JSP scripts, and a vanilla *web.xml* document. Here is the HTML form for a user to enter a temperature to be converted:

```html
<html><body>
  <form method = 'post' action = 'temp_convert.jsp'>
    Temperature to convert: <input type = 'text' name = 'temperature'><br/><hr/>
    <input type = 'submit' value = ' Click to submit '/>
  </form>
</body></html>
```

And here is the JSP script that invokes the TempConvert web service to do the temperature conversion:

```
<%@ page errorPage = 'error.jsp' %>
<%@ page import = 'client.TempConvert' %>
<%@ page import = 'client.TempConvertService' %>
<html><body>
<%! private float f2c, c2f, temp; %>
<%
    String temp_str = request.getParameter("temperature");
    if (temp_str != null) temp = Float.parseFloat(temp_str.trim());

    TempConvertService service =  new TempConvertService();
    TempConvert port = service.getTempConvertPort();
    f2c = port.f2C(temp);
    c2f = port.c2F(temp);
%>
<p><%= this.temp %>F = <%= this.f2c %>C</p>
<p><%= this.temp %>C = <%= this.c2f %>F</p>
<a href = 'index.html'>Try another</a>
</body></html>
```

The two imported classes, `client.TempConvert` and `client.TempConvertService`, are
wsimport-generated artifacts in the *WEB-INF/classes/client* subdirectory of the directory that holds the HTML document and the two JSP scripts. For completeness, here
is the *error.jsp* script:

```
<%@ page isErrorPage = "true" %>
<html>
<% response.setStatus(400); %>
<body>
<h2><%= exception.toString() %></h2>
<p>Bad data: please try again.</p>
<p><a href = "index.html">Return to home page</a></p>
</body></html>
```

The *web.xml* deployment document is also short:

```
<?xml version = '1.0' encoding = 'ISO-8859-1'?>
<web-app xmlns = 'http://java.sun.com/xml/ns/javaee'
         xmlns:xsi = 'http://www.w3.org/2001/XMLSchema-instance'
         xsi:schemaLocation = 'http://java.sun.com/xml/ns/javaee/web-app_2_5.xsd'
         version = '2.5'>
    <description>JSP frontend to TempConvert service</description>
    <error-page>
        <exception-type>java.lang.NumberFormatException</exception-type>
        <location>/error.jsp</location>
    </error-page>
    <welcome-file-list>
        <welcome-file>index.html</welcome-file>
    </welcome-file-list>
</web-app>
```

The HTML page and the JSP scripts together with the compiled classes and *web.xml* are then put into a WAR file:

```
% jar cvf tcJSP.war index.html *.jsp WEB-INF
```

for deployment in the GlassFish *autodeploy* subdirectory. For testing, a browser can be opened to the URL *http://localhost:8081/tcJSP*.

A @WebService As an EJB

The EJB container is programmer-friendly in handling issues such as thread safety, instance pooling, and transaction-guarded persistence. This section illustrates how a @WebService can take advantage of these benefits by being implemented as a *stateless* Session EJB. The example is in two steps. The first step quickly covers the details of deploying a @WebService as a stateless Session EJB. The second step adds database persistence by introducing an @Entity into the application. The example is kept simple so that the EJB and @Entity details stand out.

Implementation As a Stateless Session EJB

Here is the SEI for the FibEJB service to be implemented as a stateless Session EJB:

```
package ch06.ejb;

import java.util.List;
import javax.ejb.Stateless;
import javax.jws.WebService;
import javax.jws.WebMethod;

@Stateless
@WebService
public interface Fib {
    @WebMethod int fib(int n);
    @WebMethod List getFibs();
}
```

The main change is that the annotation @Stateless occurs. The same annotation occurs in the SIB:

```
package ch06.ejb;

import java.util.List;
import java.util.ArrayList;
import javax.ejb.Stateless;
import javax.jws.WebService;

@Stateless
@WebService(endpointInterface = "ch06.ejb.Fib")
public class FibEJB implements Fib {
    public int fib(int n) {
        int fib = 1, prev = 0;
```

```
            for (int i = 2; i <= n; i++) {
                int temp = fib;
                fib += prev;
                prev = temp;
            }
            return fib;
        }
        public List getFibs() { return new ArrayList(); } // for now, empty list
    }
```

In the spirit of a *stateless* Session EJB, the implementing class consists of self-contained methods that do not rely on any instance fields. This first version of the `getFibs` operation returns the empty list, but the next version returns the rows in a database table.

The packaging of the EJB implementation of the service differs from the standard WAR packaging. To begin, all of the required classes are placed in a JAR file whose name is arbitrary. In this case, the command:

```
% jar cvf rc.jar ch06/ejb/*.class
```

creates the file. There is no configuration document; hence, GlassFish will generate one automatically. This JAR file is then enclosed in another:

```
% jar cvf fib.ear rc.jar
```

The *EAR* (Enterprise ARchive) extension is traditional for EJBs. GlassFish expects that a deployed EAR file contains at least one EJB or `@Entity` POJO, although a production-level EAR file typically contains several EJBs. In any case, an EAR file holds one JAR file per EJB and may hold WAR files as well. The entire EAR file constitutes a single enterprise application, with the particular JAR files therein housing the various application components. In this example, there is one component: the stateless Session EJB in its own JAR file.

Once the EAR file is deployed in the usual way by being copied to *domains/domain1/ autodeploy* directory, the administrative console can be used to inspect the application, which is listed appropriately under the WebServices section in the console. GlassFish automatically generates various deployment artifacts, including the WSDL:

```
<?xml version="1.0" encoding="UTF-8"?>
<!-- Published by JAX-WS RI at http://jax-ws.dev.java.net.
     RI's version is JAX-WS RI 2.1.3.1-hudson-417-SNAPSHOT. -->
<!-- Generated by JAX-WS RI at http://jax-ws.dev.java.net.
     RI's version is JAX-WS RI 2.1.3.1-hudson-417-SNAPSHOT. -->
<definitions
  xmlns:wsu=
  "http://docs.oasis-open.org/wss/2004/01/oasis-200401-wss-wssecurity-utility-1.0.xsd"
  xmlns:soap="http://schemas.xmlsoap.org/wsdl/soap/" xmlns:tns="http://ejb.ch06/"
  xmlns:xsd="http://www.w3.org/2001/XMLSchema"
  xmlns="http://schemas.xmlsoap.org/wsdl/"
  targetNamespace="http://ejb.ch06/"
  name="FibEJBService">
```

```
<ns1:Policy xmlns:ns1="http://www.w3.org/ns/ws-policy"
            wsu:Id="FibEJBPortBinding_getFibs_WSAT_Policy">
<ns1:ExactlyOne>
<ns1:All>
<ns2:ATAlwaysCapability xmlns:ns2="http://schemas.xmlsoap.org/ws/2004/10/wsat">
</ns2:ATAlwaysCapability>
<ns3:ATAssertion xmlns:ns4="http://schemas.xmlsoap.org/ws/2002/12/policy"
    xmlns:ns3="http://schemas.xmlsoap.org/ws/2004/10/wsat"
    ns1:Optional="true"
    ns4:Optional="true">
</ns3:ATAssertion>
</ns1:All>
</ns1:ExactlyOne>
</ns1:Policy>
<ns5:Policy xmlns:ns5="http://www.w3.org/ns/ws-policy"
            wsu:Id="FibEJBPortBinding_fib_WSAT_Policy">
<ns5:ExactlyOne>
<ns5:All>
<ns6:ATAlwaysCapability xmlns:ns6="http://schemas.xmlsoap.org/ws/2004/10/wsat">
</ns6:ATAlwaysCapability>
<ns7:ATAssertion xmlns:ns8="http://schemas.xmlsoap.org/ws/2002/12/policy"
                 xmlns:ns7="http://schemas.xmlsoap.org/ws/2004/10/wsat"
                 ns5:Optional="true" ns8:Optional="true">
</ns7:ATAssertion>
</ns5:All>
</ns5:ExactlyOne>
</ns5:Policy>

<types>
<xsd:schema>
<xsd:import namespace="http://ejb.ch06/"
            schemaLocation="http://localhost:8080/FibEJBService/FibEJB?xsd=1">
</xsd:import>
</xsd:schema>
</types>
<message name="fib">
<part name="parameters" element="tns:fib"></part>
</message>
<message name="fibResponse">
<part name="parameters" element="tns:fibResponse"></part>
</message>
<message name="getFibs">
<part name="parameters" element="tns:getFibs"></part>
</message>
<message name="getFibsResponse">
<part name="parameters" element="tns:getFibsResponse"></part>
</message>

<portType name="Fib">
<operation name="fib">
<ns9:PolicyReference xmlns:ns9="http://www.w3.org/ns/ws-policy"
            URI="#FibEJBPortBinding_fib_WSAT_Policy"></ns9:PolicyReference>
<input message="tns:fib"></input>
<output message="tns:fibResponse"></output>
</operation>
```

```
<operation name="getFibs">
<ns10:PolicyReference xmlns:ns10="http://www.w3.org/ns/ws-policy"
                URI="#FibEJBPortBinding_getFibs_WSAT_Policy"></ns10:PolicyReference>
<input message="tns:getFibs"></input>
<output message="tns:getFibsResponse"></output>
</operation>
</portType>
<binding name="FibEJBPortBinding" type="tns:Fib">
<soap:binding transport="http://schemas.xmlsoap.org/soap/http"
                style="document">
</soap:binding>
<operation name="fib">
<ns11:PolicyReference xmlns:ns11="http://www.w3.org/ns/ws-policy"
                URI="#FibEJBPortBinding_fib_WSAT_Policy"></ns11:PolicyReference>
<soap:operation soapAction=""></soap:operation>

<input>
<soap:body use="literal"></soap:body>
</input>
<output>
<soap:body use="literal"></soap:body>
</output>
</operation>
<operation name="getFibs">
<ns12:PolicyReference xmlns:ns12="http://www.w3.org/ns/ws-policy"
                URI="#FibEJBPortBinding_getFibs_WSAT_Policy"></ns12:PolicyReference>
<soap:operation soapAction=""></soap:operation>
<input>
<soap:body use="literal"></soap:body>
</input>
<output>
<soap:body use="literal"></soap:body>
</output>
</operation>
</binding>
<service name="FibEJBService">
<port name="FibEJBPort" binding="tns:FibEJBPortBinding">
<soap:address location="http://localhost:8080/FibEJBService/FibEJB"></soap:address>
</port>
</service>
</definitions>
```

This WSDL is unlike any of the previous ones in that it contains a section at the top on
WS-Policy, which is a language for describing web service capabilities and client requirements. GlassFish includes the current Metro release, which supports WS-Policy
and related WS-* initiatives that promote web services interoperability: WSIT in Metro
speak. The policy templates are laid out at the top of the WSDL and then referenced
throughout the rest of the document. However, all of the policy templates are marked
as merely optional. The underlying idea is that a web service and its potential consumers
should be able to articulate policies about security and other service-related matters.
For example, there is a set of policies on reliable messaging that belongs to the WS-
ReliableMessaging specification. Under this specification, a WSDL could advertise that
any message is to be sent exactly once and, therefore, must never be duplicated. If an

intermediary along the path from the sender to the receiver encounters a problem in delivering a message, the appropriate fault must be raised and the message must not be resent. Another WS-ReliableMessaging policy stipulates that messages will be received in the same order in which they are sent, a feature that the TCP infrastructure of HTTP/HTTPS already guarantees. Yet because SOAP-based web services are designed to be transport neutral, there is a specification on reliable messaging, several on security, and so on.

The Endpoint URL for an EBJ-Based Service

The endpoint URL for the EJB-based web service differs from WAR examples seen so far. For one thing, the name of the EAR file does *not* occur in the path as does the name of a WAR file. There are two pieces in the path section of the URL: the first is the SIB name with `Service` appended, giving `FibEJBService` in this example; the second is the SIB name, in this case `FibEJB`. The combination is shown in the URL below for the sample Perl client:

```
#!/usr/bin/perl -w
use SOAP::Lite;
use strict;
my $url = 'http://localhost:8081/FibEJBService/FibEJB?wsdl';
my $service = SOAP::Lite->service($url);
print $service->fib(7), "\n";
```

For convenience and simplicity, this example uses only defaults. For example, the EJB is deployed with the name `FibEJBService` and the path `/FibEJBService/FibEJB`. These defaults can be overridden by adding attributes in an annotation or by entries in a deployment descriptor. The good news is that there are defaults.

The next step is to add a database table as the persistence store and an `@Entity` to automate the interaction between the web service and the database. This step requires only modest additions to the service.

Database Support Through an @Entity

The SEI interface `Fib` stays the same, but the SIB `FibEJB` changes to the following:

```
package ch06.ejb;
import java.util.List;
import java.util.ArrayList;
import javax.ejb.Stateless;
import javax.jws.WebService;
import javax.persistence.PersistenceContext;
import javax.persistence.EntityManager;
import javax.persistence.Query;
@Stateless
@WebService(endpointInterface = "ch06.ejb.Fib")
public class FibEJB implements Fib {
    @PersistenceContext(unitName = "FibServicePU")
    private EntityManager em;
```

```
public int fib(int n) {
    // Computed already? If so, return.
    FibNum fn = em.find(FibNum.class, n); // read from database
    if (fn != null) return fn.getF();

    int f = compute_fib(Math.abs(n));
    fn = new FibNum();
    fn.setN(n);
    fn.setF(f);
    em.persist(fn);                        // write to database
    return f;
}

public List getFibs() {
    Query query = em.createNativeQuery("select * from FibNum");
    // fib_nums is a list of pairs: N and Fibonacci(N)
    List fib_nums = query.getResultList(); // read from database
    List results = new ArrayList();
    for (Object next : fib_nums) {
        List list = (List) next;
        for (Object n : list) results.add(n);
    }
    return results; // N, fib(N), K, fib(K),...
}

private int compute_fib(int n) {
    int fib = 1, prev = 0;
    for (int i = 2; i <= n; i++) {
        int temp = fib;
        fib += prev;
        prev = temp;
    }
    return fib;
}
}
```

On a client request for a particular Fibonacci number, the service now does a find operation against the database. If the value is not stored in the database, then the value is computed and stored there. If the client requests a list of all Fibonacci numbers computed so far, the service executes a query against the database to get the list, which is returned. This overview now can be fleshed out with details.

The FibEJB SIB uses dependency injection to get a reference to an EntityManager that, as the name indicates, manages any @Entity within a given PersistenceContext. The @Entity in this case is a FibNum:

```
package ch06.ejb;

import javax.persistence.Entity;
import javax.persistence.Id;
import javax.persistence.Column;
import java.io.Serializable;
```

```
// A FibNum is a pair: an integer N and its Fibonacci value.
@Entity
public class FibNum implements Serializable {
    private int n;
    private int f;

    public FibNum() { }

    @Id
    public int getN() { return n; }
    public void setN(int n) { this.n = n; }

    public int getF() { return f; }
    public void setF(int f) { this.f = f; }
}
```

A FibNum has two properties, one for an integer N and another for the Fibonacci value
of N. In the corresponding database table, the integer N serves as the primary key; that
is, as the @Id. How the database is created will be explained shortly. First, though, a bit
more on the EntityManager is in order.

By default in the EJB container, a PersistenceContext includes transaction support
through a *JTA* (Java Transaction API) manager; hence, transactions are automatic because the container introduces and manages them. Again by default in this example,
the scope of the transaction is a single *read* or *write* operation against the database. The
method fib invokes the EntityManager method find to retrieve a FibNum from the database, with each FibNum as one row in the table. If a particular FibNum is not in the table,
then a new @Entity instance is created and persisted with a call to the EntityManager
method persist.

The methods encapsulated in an EntityManager such as find and persist are not inherently thread-safe. These methods become thread-safe in this example because the
EntityManager instance is a field in an EJB and, therefore, enjoys the thread safety that
the EJB container bestows.

The Persistence Configuration File

Nothing in the Web service code or the support code explicitly references the database
or the table therein. These details are handled through a configuration document,
META-INF/persistence.xml:

```
<persistence>
    <persistence-unit name="FibServicePU" transaction-type="JTA">
        <description>
            This unit manages Fibonacci number persistence.
        </description>
        <jta-data-source>jdbc/__default</jta-data-source>
        <properties>
          <!--Use the java2db feature -->
          <property name="toplink.ddl-generation" value="drop-and-create-tables"/>
          <!-- Generate the sql specific to Derby database -->
```

```
                <property name="toplink.platform.class.name"
                    value="oracle.toplink.essentials.platform.database.DerbyPlatform"/>
            </properties>
            <class>ch06.ejb.FibNum</class>
        </persistence-unit>
    </persistence>
```

Each JAR file with an @Entity should have a *META-INF* directory with a file named *persistence.xml*. The lines in bold need explanation.

GlassFish comes with several JDBC resources, including the one with the JNDI name jdbc/__default used here. The administrative console has a utility for creating new databases. As noted earlier, GlassFish also ships with the Apache Derby database system, which in turn includes the TopLink utility called *java2db* that generates a table schema from a Java class, in this case the class FibNum. Recall that FibNum is annotated as an @Entity and has two properties, one for an integer N (the primary key) and another for the Fibonacci value of N. Under the hood the *java2db* utility generates the corresponding database table with two columns, one for N and another for its Fibonacci value. If desired, the names of the database table and the columns can be specified with the @Table and @Column annotations, respectively.

In the FibEJB service, the @WebMethod named fib invokes the find method whose argument is the primary key for the desired FibNum pair. By contrast, the method with the name getFibs relies on a query to retrieve all of the rows in the database. These rows are returned as a list of pair lists, in effect as ((1,1), (2,1), (3,2),...). The getFibs method uses a native as opposed to a *JPQL* (Java Persistence Query Language) query to retrieve the rows and then creates a simple list of integers to return to the requester. The returned list looks like (1, 1, 2, 1, 3, 2,...), with N and its Fibonacci value as adjacent values.

Here is a sample client against the FibEJB service:

```
import clientEJB.FibEJBService;
import clientEJB.Fib;
import java.util.List;

class ClientEJB {
    public static void main(String args[ ]) {
        FibEJBService service = new FibEJBService();
        Fib port = service.getFibEJBPort();

        final int n=8;
        for (int i=1; i<n; i++) System.out.println("Fib(" + i + ") == "+port.fib(i));

        List fibs = port.getFibs();
        for (Object next : fibs) System.out.println(next);
    }
}
```

The output is:

```
Fib(1) == 1
Fib(2) == 1
Fib(3) == 2
```

```
Fib(4) == 3
Fib(5) == 5
Fib(6) == 8
Fib(7) == 13
1
1
2
1
3
2
4
3
5
5
6
8
7
13
```

The EJB Deployment Descriptor

The `FibEJB` service is implemented as an EJB and packaged in an EAR file without a deployment descriptor (DD), which is an XML document in the file *ejb-jar.xml*. In an EAR file, each EJB occurs in its own JAR file that optionally has a DD for the EJB. Here is the DD that GlassFish generates for `FibEJB`:

```
<?xml version="1.0" encoding="UTF-8" standalone="no"?>
<ejb-jar xmlns="http://java.sun.com/xml/ns/javaee"
         xmlns:xsi="http://www.w3.org/2001/XMLSchema-instance"
         metadata-complete="true" version="3.0"
         xsi:schemaLocation="http://java.sun.com/xml/ns/javaee
                             http://java.sun.com/xml/ns/javaee/ejb-jar_3_0.xsd">
  <enterprise-beans>
    <session>
      <display-name>Fib</display-name>
      <ejb-name>Fib</ejb-name>
      <ejb-class>ch06.ejb.Fib</ejb-class>
      <session-type>Stateless</session-type>
      <transaction-type>Container</transaction-type>
      <security-identity>
        <use-caller-identity/>
      </security-identity>
    </session>
    <session>
      <display-name>FibEJB</display-name>
      <ejb-name>FibEJB</ejb-name>
      <business-local>ch06.ejb.Fib</business-local>
      <service-endpoint>ch06.ejb.Fib</service-endpoint>
      <ejb-class>ch06.ejb.FibEJB</ejb-class>
      <session-type>Stateless</session-type>
      <transaction-type>Container</transaction-type>
      <persistence-context-ref>
        <persistence-context-ref-name>ch06.ejb.FibEJB/em
        </persistence-context-ref-name>
```

```
        <persistence-unit-name>FibServicePU</persistence-unit-name>
        <persistence-context-type>Transaction</persistence-context-type>
        <injection-target>
          <injection-target-class>ch06.ejb.FibEJB</injection-target-class>
          <injection-target-name>em</injection-target-name>
        </injection-target>
      </persistence-context-ref>
      <security-identity>
        <use-caller-identity/>
      </security-identity>
    </session>
  </enterprise-beans>
</ejb-jar>
```

There are two `<session>` sections, one at the top for the SEI `Fib` and another at the bottom for the SIB `FibEJB`. Each section indicates that the Session EJB is deployed as stateless with container-managed transactions. The SIB section provides additional details about persistence, including the reference `em` that refers to the `EntityManager` used in `FibEJB`. The key point is that the persistence context is transaction-based, with the EJB container responsible for transaction management.

Servlet and EJB Implementations of Web Services

GlassFish allows web services to be deployed in either WAR or EAR files. In the first case, GlassFish characterizes the implementation as *servlet-based* because, as seen in the examples with standalone Tomcat, a `WSServlet` instance intercepts that web service requests and passes these on to the web service packaged in the WAR file. In the second case, the web service is a stateless Session EJB that the EJB container manages. In the servlet implementation, the web container handles requests against the web service; in the EJB implementation, the EJB container handles requests against the web service. A given web service thus can be implemented in two different ways. At issue, then, are the tradeoffs between the two implementations of a web service.

The main advantage of the EJB implementation is that the EJB container provides more services than does the servlet container. For one thing, a web service implemented as an EJB is thereby thread-safe, whereas the operationally identical service implemented as a POJO is not thread-safe. In the `FibEJB` example, the web operation `fib` does a *read* and a possible *write* against the database, and the operation `getFibs` does a *read*. The EJB container also automatically wraps these database operations in transactions, which remain transparent. The EJB container manages the persistence so that the application does not have to do so. In a servlet-based or `Endpoint`-based service, the application would take on these responsibilities.

Prior to EJB 3.0, a strong case could be made that EJBs in general, including even Session EJBs, were simply too much work for too little reward. The case is no longer so strong. As the `FibEJB` example shows, the EJB implementation of a SOAP-based service requires only one additional annotation, `@Stateless`, and the packaging of an EJB-based service in an EAR file is arguably no harder than packaging the same service in a WAR file with its *WEB-INF/classes* structure.

Java Web Services and Java Message Service

GlassFish, like any full-feature application server, includes a JMS provider that supports publish/subscribe (topic-based) and point-to-point (queue-based) messaging. Topics and queues are persistent stores. For example, one application can send a message to a JMS queue, and another application, at some later time, can retrieve the message from the queue. However, the JMS does not specify how long an implementation needs to persist messages. This section illustrates the basics of JWS/JMS interaction with two web services. The MsgSender service has a send operation that inserts a message in a queue, and the MsgReceiver service has a receive operation that retrieves a message from the front of the queue (see Figure 6-2).

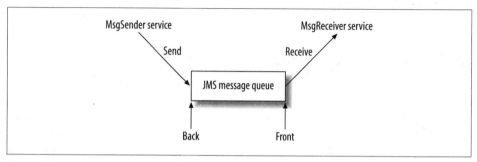

Figure 6-2. Integrating JAX-WS and JMS

Here is the source code for the MsgSender service:

```
package ch06.jms;

import javax.jws.WebService;
import javax.jws.WebMethod;
import javax.jms.Queue;
import javax.jms.Session;
import javax.jms.TextMessage;
import javax.jms.QueueConnectionFactory;
import javax.jms.QueueConnection;
import javax.jms.Session;
import javax.jms.JMSException;
import javax.annotation.Resource;

// A web service that sends a message to a queue.
@WebService
public class MsgSender {
    // name and mappedName can differ; mappedName is the JNDI lookup name
    @Resource(name="qcf", mappedName="qcf")
    private QueueConnectionFactory qf;
```

```
@Resource(name="jmsQ", mappedName="jmsQ")
private Queue queue;

private QueueConnection conn;

@WebMethod
public void send(String msg) {
    try {
        if (conn == null) conn = (QueueConnection) qf.createConnection();
        Session session =
            conn.createSession(false, // no transaction support
                                Session.AUTO_ACKNOWLEDGE);

        // Wrap the string in a TextMessage for sending.
        TextMessage tmsg = session.createTextMessage(msg);
        session.createProducer(queue).send(tmsg);
        session.close();
    }
    catch(JMSException e) { throw new RuntimeException(e); }
    }
}
```

Setting Up GlassFish Queues and Topics

The *asadmin* utility in the *AS_HOME/bin* directory can be used to check whether the JMS provider is up and running. The command is:

```
% asadmin jms-ping
```

A status report such as:

```
JMS Ping Status = RUNNING
Command jms-ping executed successfully.
```

indicates whether the JMS provider is running. The JMS provider should start automatically when GlassFish starts, although it is possible to configure GlassFish so that the JMS provider must be started separately.

The GlassFish administrative console has a `Resources` tab and a `JMS Resources` subtab for creating named `Connection Factories` for a topic or a queue and named `Destination Resources`, which are topics or queues. In the code that follows, the name of a connection factory and the name of a queue are used with `@Resource` annotations.

The `MsgSender` service relies on dependency injection, enabled through the `@Resource` annotation, to get the connection factory and the queue created through the GlassFish administrative console. Sending the message requires a connection to the JMS provider, a session with the provider, and a message producer that encapsulates the `send` method. JMS has several message types, including the `TextMessage` type used here. Once the message is sent to the queue, the session is closed but the connection to the JMS provider remains open for subsequent operations against the queue.

The `MsgReceiver` service uses the same dependency injection code. The setup is also similar except that a `QueueSession` replaces a generic `Session` object. To show some of the variety in the JMS API, a `QueueReceiver` is introduced. The invocation of the `QueueConnection` method `start` completes the setup phase and enables the message at the front of queue, if there is one, to be retrieved. Here is the code:

```java
package ch06.jms;

import java.util.List;
import java.util.ArrayList;
import javax.jws.WebService;
import javax.jws.WebMethod;
import javax.jms.Message;
import javax.jms.TextMessage;
import javax.jms.Session;
import javax.jms.Queue;
import javax.jms.QueueSession;
import javax.jms.QueueConnectionFactory;
import javax.jms.QueueConnection;
import javax.jms.QueueSession;
import javax.jms.QueueReceiver;
import javax.jms.JMSException;
import javax.annotation.Resource;

@WebService
public class MsgReceiver {
    // name and mappedName can differ
    @Resource(name="qcf", mappedName="qcf")
    private QueueConnectionFactory qf;
    @Resource(name="jmsQ", mappedName="jmsQ")
    private Queue queue;

    private QueueConnection conn;

    @WebMethod
    public String receive() {
        String cliche = null;
        try {
            if (conn == null) conn = qf.createQueueConnection();
            QueueSession session =
                conn.createQueueSession(false,
                                        Session.AUTO_ACKNOWLEDGE);
            QueueReceiver receiver = session.createReceiver(queue);
            conn.start();

            Message msg = receiver.receiveNoWait();
            if (msg != null && msg instanceof TextMessage ) {
                TextMessage tmsg = (TextMessage) msg;
                cliche = tmsg.getText().trim();
            }
        }
        catch(JMSException e) { throw new RuntimeException(e); }
        return cliche;
    }
}
```

There are three JMS methods for retrieving a message from a queue. This example uses `receiveNoWait`, which behaves as the name indicates: the method returns the message at the front of the queue, if there is one, but immediately returns `null`, if the queue is empty. There is also an overloaded `receive`s method. The no-argument version blocks until there is a message to retrieve, whereas the one-argument version blocks only n milliseconds, for n > 0, if the queue is currently empty.

JMS also has a `MessageListener` interface, which declares an `onMessage(Message)` method. A message consumer such a `QueueReceiver` has a `setMessageListener` method whose argument is an instance of a class that implements the `MessageListener` interface. The `onMessage` method then behaves as a callback that is invoked whenever a message arrives for consumption.

Within the Java community, there is ongoing interest in the interaction between JWS and JMS. For example, there is a third-party initiative to provide a SOAP over JMS binding. Even without this initiative, of course, a SOAP envelope could be encapsulated as the body of a JMS message; or a JMS message could be encapsulated as the cargo of a SOAP envelope. There also are packages for converting between JMS and SOAP. The example in this section emphasizes the store-and-forward capabilities of JMS and shows how these might be integrated with SOAP-based web services.

WS-Security Under GlassFish

The `Echo` service example in Chapter 5 shows that WS-Security can be used with `Endpoint` by using the Metro packages. WS-Security is easier under GlassFish precisely because the current Metro release is part of the GlassFish distribution. This section illustrates the point.

Making an HTTP Dump Automatic Under GlassFish

In any GlassFish deployment domain such as *domain1*, there is a subdirectory *config* with various files, including *domain.xml*. This configuration file has a section for JVM options, each of which has the start tag `<jvm-options>`. To enable an automatic HTTP dump of the SOAP messages and their security artifacts, two JVM options should be added:

```
<jvm-options>
  -Dcom.sun.xml.ws.transport.http.HttpAdapter.dump=true
</jvm-options>
<jvm-options>
  -Dcom.sun.xml.ws.transport.http.client.HttpTransportPipe.dump=true
</jvm-options>
```

If the application server is already running, it must be restarted for the new configuration to take effect. The dump is to the *domains/domain1/logs/server.log* file.

The first example focuses on peer authentication using digital certificates.

Mutual Challenge with Digital Certificates

In a typical browser-based web application, the browser challenges the web server to authenticate itself when the browser tries to establish an HTTPS connection with the server. As discussed earlier in Chapter 5, the web server typically does not challenge the client. For instance, the default behavior of Tomcat is *not* to challenge a client. GlassFish as a whole has the same default behavior. In the administrative console, under the Security tab, this default behavior can be changed so that GlassFish, including embedded Tomcat, challenges the client automatically. The approach in this example is to force the challenge at the application level rather than for the JAS as a whole. In any case, a challenged party may respond with digital certificates that vouch for its identity. This section illustrates mutual challenge using digital certificates, a scenario that the acronym *MCS* (Mutual Certificates Security) captures.

Recall the scenario in which Alice wants to send a message securely to Bob, with each of the two relying upon digital certificates to authenticate the identity of the other. Alice's *keystore* holds her own digital certificates, which she can send to Bob, whereas Alice's *truststore* holds digital certificates that establish her trust in Bob. Alice's truststore may contain earlier verified certificates from Bob or certificates from a CA that vouch for Bob's certificates. For convenience in testing, the keystore and truststore could be the same file.

There are two ways to do an MCS application. One way is to rely directly on HTTPS, which incorporates mutual challenge as part of the setup phase in establishing a secure communications channel between the client and the web service. On the client side, CLIENT-CERT can be used instead of BASIC or DIGEST authentication so that the client, like the server, presents a digital certificate to vouch for its identity. The second way is to be transport-agnostic and rely instead on mutual challenge with Metro's WSIT support. (Recall that the *I* in *WSIT* is for Interoperability, which includes transport neutrality.) Each way is illustrated with an example. The HTTPS example implements the web service with an EJB endpoint, whereas the WSIT example implements the web service with a servlet endpoint.

MCS Under HTTPS

The source code for the web service is neutral with respect to security and so could be deployed under either HTTP or HTTPS:

```
package ch06.sslWS;

import javax.ejb.Stateless;
import javax.jws.WebService;

@Stateless
@WebService
public class EchoSSL {
    public String echo(String msg) { return "Echoing: " + msg; }
}
```

The EchoSSL service is first deployed without any security to underscore that a secured version does not require changes to the source code. The *wsimport* artifacts can be generated in the usual way from the unsecured version or, using *wsgen*, from the secured version. In any case, GlassFish awaits HTTPS connections on port 8181.

Security for the EchoSSL service is imposed with a GlassFish-specific configuration file, *sun-ejb-jar.xml*:

```
<?xml version="1.0" encoding="UTF-8"?>
<!DOCTYPE sun-ejb-jar PUBLIC '-//Sun Microsystems, Inc.//
         DTD Sun ONE Application Server 8.0 EJB 2.1//EN'
         'http://www.sun.com/software/sunone/appserver/
         dtds/sun-ejb-jar_2_1-0.dtd'>
<sun-ejb-jar>
  <enterprise-beans>
    <ejb>
      <ejb-name>EchoSSL</ejb-name>
      <webservice-endpoint>
        <port-component-name>EchoSSL</port-component-name>
        <login-config>
          <auth-method>CLIENT-CERT</auth-method>
          <realm>certificate</realm>
        </login-config>
        <transport-guarantee>CONFIDENTIAL</transport-guarantee>
      </webservice-endpoint>
    </ejb>
  </enterprise-beans>
</sun-ejb-jar>
```

The security section of this configuration file is essentially the same as the security-constraint section of a *web.xml* file. Two sections of interest are the login-config section, which sets the authentication to CLIENT-CERT, and the transport-guarantee section, which calls for data confidentiality across the communications channel. This configuration file is included in the JAR file that holds the compiled web service, in this example *echoSSL.jar*. The name of the JAR file is arbitrary, but the file *sun-ejb-jar.xml* must be in the *META-INF* directory within the JAR file. Here, then, is the layout of the JAR file:

```
META-INF/sun-ejb-jar.xml
ch06/sslWS/EchoSSL.class
```

The JAR file with the configuration document and the stateless EJB is then put in an EAR file as usual. As earlier examples show, an EAR file does not require a configuration document, but the standard configuration document is named *application.xml*. Here, for illustration, is the configuration document used in this example:

```
<?xml version="1.0" encoding="UTF-8"?>
<!DOCTYPE application PUBLIC "-//Sun Micro., Inc.//DTD J2EE Application 1.3//EN"
          "http://java.sun.com/dtd/application_1_3.dtd">
<application>
  <display-name>EchoSSL</display-name>
  <module>
    <ejb>echoSSL.jar</ejb>
```

```
        </module>
    </application>
```

The *META-INF/application.xml* file is a manifest that lists each JAR file within the EAR file as a `module` that belongs to the application as a whole. In this case, then, the application consists of the stateless EJB in *echoSSL.jar*. Once created, the EAR file is deployed as usual by being copied to *domains/domain1/autodeploy*.

The client code, like the service code, provides no hint that the client must connect over HTTPS:

```
import clientSSL.EchoSSLService;
import clientSSL.EchoSSL;

class EchoSSLClient {
    public static void main(String[ ] args) {
        try {
            EchoSSLService service = new EchoSSLService();
            EchoSSL port = service.getEchoSSLPort();
            System.out.println(port.echo("Goodbye, cruel world!"));
        }
        catch(Exception e) { System.err.println(e); }
    }
}
```

Although the client code is straightforward, its execution requires information about the client's keystore and truststore. GlassFish itself provides a default keystore and truststore for development mode. The keystore and the truststore for *domain1* are in *domains/domain1/config* and are named *keystore.jks* and *cacerts.jks*, respectively. The keystore contains a self-signed certificate. In production mode, of course, a CA-signed certificate presumably would be used. To make testing as easy as possible, the client is simply invoked with the GlassFish keystore and truststore. Here is the execution command with the output:

```
% java -Djavax.net.ssl.trustStore=cacerts.jks \
       -Djavax.net.ssl.trustStorePassword=changeit \
       -Djavax.net.ssl.keyStore=keystore.jks \
       -Djavax.net.ssl.keyStorePassword=changeit EchoSSLClient
Echoing: Goodbye, cruel world!
```

The underlying SOAP messages are also indistinguishable from those sent over HTTP:

```
<?xml version="1.0" ?>
<S:Envelope xmlns:S="http://schemas.xmlsoap.org/soap/envelope/">
  <S:Body>
    <ns2:echo xmlns:ns2="http://sslWS.ch06/">
      <arg0>Goodbye, cruel world!</arg0>
    </ns2:echo>
  </S:Body>
</S:Envelope>
```

The security does not show up at the SOAP level because the security is provided at the transport level; that is, through HTTPS. Only the configuration documents indicate that MCS is in play. In the WSIT example that follows, the SOAP messages will change dramatically because the security occurs at the SOAP level.

MCS Under WSIT

For variety, the MCS client is now a servlet that receives a text input from a browser or the equivalent and invokes the EchoService method echo with this string as the argument. The servlet responds to the browser with the string returned from the echo method. HTTP comes into play only in connecting the client's browser to the servlet container. Within the servlet container, the EchoService and the EchoClient servlet do exchange SOAP messages, but not over HTTP (see Figure 6-3).

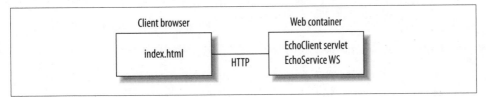

Figure 6-3. Mutual challenge under WS-Security

WSIT MCS is introduced into the application through configuration files. These files are sufficiently complicated that it may be best to generate them with a tool. The NetBeans IDE suits the purpose by producing the configuration files and the related Ant scripts that handle compilation, packaging, and deployment. It should be emphasized again, however, that this and any other WSIT application can be done without the NetBeans IDE.

The @WebService is now servlet rather than EJB-based, although this change by itself plays no role in the MCS. Here is the source:

```
package ch06.mcs;

import javax.jws.WebService;
import javax.jws.WebMethod;

@WebService
public class Echo {
    @WebMethod
    public String echo(String msg) { return "Echoing: " + msg; }
}
```

The client is a servlet that uses the usual *wsimport*-generated artifacts:

```
package ch06.mcs.client;

import java.io.PrintWriter;
import java.io.IOException;
import java.io.Closeable;
import javax.annotation.Resource;
```

```java
import javax.servlet.ServletException;
import javax.servlet.http.HttpServlet;
import javax.servlet.http.HttpServletRequest;
import javax.servlet.http.HttpServletResponse;
import javax.xml.ws.WebServiceRef;
import javax.xml.ws.soap.SOAPFaultException;

public class EchoClient extends HttpServlet {
    @WebServiceRef(wsdlLocation = "http://localhost:8080/echoMCS/EchoService?wsdl")
    public EchoService service;

    public void doGet(HttpServletRequest req, HttpServletResponse res) {
        res.setContentType("text/html;charset=UTF-8");
        PrintWriter out = null;
        try {
            out = res.getWriter();
            out.println("<h3>EchoServlet: " + req.getContextPath() + "</h3>");

            // Get the port reference and invoke echo method.
            Echo port = service.getEchoPort();
            String result = port.echo(req.getParameter("msg"));

            // If there's no SOAP fault so far, authentication worked.
            out.println("<h3>Authentication OK</h3><br/>");
            out.println(result);
            out.flush();
            ((Closeable) port).close(); // close connection to service
        }
        catch (SOAPFaultException e) {
            out.println("Authentication failure: " + e);
        }
        catch (Exception e) { out.println(e); }
        finally { out.close(); }
    }

    public void doPost(HttpServletRequest req, HttpServletResponse res) {
        try {
            this.doGet(req, res); // shouldn't happen but just in case :)
        }
        catch(Exception e) { throw new RuntimeException("doPost"); }
    }
}
```

The servlet container honors the @WebServiceRef annotation, shown in bold, and initializes the EchoService reference service. Once the web service operation echo has been invoked and the response sent back to the browser, the EchoClient closes the port to signal to the container that further access to the web service is not required on this request.

The EchoClient servlet is invoked on each submission of the simple HTML form in which a client enters a phrase to be echoed. For completeness, here is the form:

```html
<html>
    <head/>
    <body>
        <h3>EchoService Client</h3>
        <form action = 'EchoServlet' method = 'GET'><br/>
            <input type = 'text' name = 'msg'/><br/>
            <p><input type = 'submit' value = 'Send message'/></p>
        </form>
    </body>
</html>
```

Nothing so far indicates that MCS is in play. It is the configuration documents alone that introduce MCS into the application. Each side, the Echo web service and the EchoClient servlet, has a small *sun-web.xml* that provides the container with the context root, which is the name of the deployed WAR file. Following is the document for the EchoClient:

```xml
<?xml version="1.0" encoding="UTF-8"?>
<!DOCTYPE sun-web-app PUBLIC
        "-//Sun Microsystems, Inc.//DTD Application Server 9.0 Servlet 2.5//EN"
        "http://www.sun.com/software/appserver/dtds/sun-web-app_2_5-0.dtd">
<sun-web-app error-url="">
  <context-root>/echoClient</context-root>
</sun-web-app>
```

The *sun-web.xml* documents reside in the *WEB-INF* directory of each deployed WAR file. There is a second configuration document, which is considered next.

The major MCS configuration document for the service and its client begins with *wsit-*: *wsit-ch06.mcs.Echo.xml* in the case of the web service and *wsit-client.xml* in the case of the client. Each document has the structure of a WSDL. (However, *wsit-client.xml* imports another document, which contains most of the configuration information.) On the web service side, the document resides in the *WEB-INF* directory; on the client side, the document resides in the *WEB-INF/classes/META-INF* directory. Here is the service-side *wsit-ch06.mcs.Echo.xml*:

```xml
<?xml version="1.0" encoding="UTF-8"?>
<definitions
 xmlns="http://schemas.xmlsoap.org/wsdl/"
 xmlns:wsdl="http://schemas.xmlsoap.org/wsdl/"
 xmlns:xsd="http://www.w3.org/2001/XMLSchema"
 xmlns:soap="http://schemas.xmlsoap.org/wsdl/soap/"
 name="EchoService"
 targetNamespace="http://mcs.ch06/"
 xmlns:tns="http://mcs.ch06/"
 xmlns:wsp="http://schemas.xmlsoap.org/ws/2004/09/policy"
 xmlns:wsu="http://docs.oasis-open.org/wss/2004/01/oasis-200401-wss-wssecurity-
            utility-1.0.xsd"
 xmlns:wsaws="http://www.w3.org/2005/08/addressing"
 xmlns:sp="http://schemas.xmlsoap.org/ws/2005/07/securitypolicy"
 xmlns:sc="http://schemas.sun.com/2006/03/wss/server"
 xmlns:wspp="http://java.sun.com/xml/ns/wsit/policy" >
    <message name="echo"/>
    <message name="echoResponse"/>
```

```
<portType name="Echo">
    <wsdl:operation name="echo">
        <wsdl:input message="tns:echo"/>
        <wsdl:output message="tns:echoResponse"/>
    </wsdl:operation>
</portType>
<binding name="EchoPortBinding" type="tns:Echo">
    <wsp:PolicyReference URI="#EchoPortBindingPolicy"/>
    <wsdl:operation name="echo">
        <wsdl:input>
            <wsp:PolicyReference URI="#EchoPortBinding_echo_Input_Policy"/>
        </wsdl:input>
        <wsdl:output>
            <wsp:PolicyReference URI="#EchoPortBinding_echo_Output_Policy"/>
        </wsdl:output>
    </wsdl:operation>
</binding>
<service name="EchoService">
    <wsdl:port name="EchoPort" binding="tns:EchoPortBinding"/>
</service>
<wsp:Policy wsu:Id="EchoPortBindingPolicy">
    <wsp:ExactlyOne>
        <wsp:All>
            <wsaws:UsingAddressing
                xmlns:wsaws="http://www.w3.org/2006/05/addressing/wsdl"/>
            <sp:SymmetricBinding>
                <wsp:Policy>
                    <sp:ProtectionToken>
                        <wsp:Policy>
                            <sp:X509Token
                                sp:IncludeToken="http://schemas.xmlsoap.org/ws/
                                        2005/07/securitypolicy/IncludeToken/Never">
                                <wsp:Policy>
                                    <sp:WssX509V3Token10/>
                                </wsp:Policy>
                            </sp:X509Token>
                        </wsp:Policy>
                    </sp:ProtectionToken>
                    <sp:Layout>
                        <wsp:Policy>
                            <sp:Strict/>
                        </wsp:Policy>
                    </sp:Layout>
                    <sp:IncludeTimestamp/>
                    <sp:OnlySignEntireHeadersAndBody/>
                    <sp:AlgorithmSuite>
                        <wsp:Policy>
                            <sp:Basic128/>
                        </wsp:Policy>
                    </sp:AlgorithmSuite>
                </wsp:Policy>
            </sp:SymmetricBinding>
            <sp:Wss11>
                <wsp:Policy>
                    <sp:MustSupportRefKeyIdentifier/>
```

```
                        <sp:MustSupportRefIssuerSerial/>
                        <sp:MustSupportRefThumbprint/>
                        <sp:MustSupportRefEncryptedKey/>
                    </wsp:Policy>
                </sp:Wss11>
                <sp:SignedSupportingTokens>
                    <wsp:Policy>
                        <sp:UsernameToken
                            sp:IncludeToken="http://schemas.xmlsoap.org/ws/2005/07/
                                    securitypolicy/IncludeToken/AlwaysToRecipient">
                            <wsp:Policy>
                                <sp:WssUsernameToken10/>
                            </wsp:Policy>
                        </sp:UsernameToken>
                    </wsp:Policy>
                </sp:SignedSupportingTokens>
                <sc:KeyStore wspp:visibility="private" alias="xws-security-server"/>
            </wsp:All>
        </wsp:ExactlyOne>
    </wsp:Policy>
    <wsp:Policy wsu:Id="EchoPortBinding_echo_Input_Policy">
        <wsp:ExactlyOne>
            <wsp:All>
                <sp:EncryptedParts>
                    <sp:Body/>
                </sp:EncryptedParts>
                <sp:SignedParts>
                    <sp:Body/>
                    <sp:Header Name="To"
                                Namespace="http://www.w3.org/2005/08/addressing"/>
                    <sp:Header Name="From"
                                Namespace="http://www.w3.org/2005/08/addressing"/>
                    <sp:Header Name="FaultTo"
                                Namespace="http://www.w3.org/2005/08/addressing"/>
                    <sp:Header Name="ReplyTo"
                                Namespace="http://www.w3.org/2005/08/addressing"/>
                    <sp:Header Name="MessageID"
                                Namespace="http://www.w3.org/2005/08/addressing"/>
                    <sp:Header Name="RelatesTo"
                                Namespace="http://www.w3.org/2005/08/addressing"/>
                    <sp:Header Name="Action"
                                Namespace="http://www.w3.org/2005/08/addressing"/>
                    <sp:Header Name="AckRequested"
                                Namespace="http://schemas.xmlsoap.org/ws/2005/02/rm"/>
                    <sp:Header Name="SequenceAcknowledgement"
                                Namespace="http://schemas.xmlsoap.org/ws/2005/02/rm"/>
                    <sp:Header Name="Sequence"
                                Namespace="http://schemas.xmlsoap.org/ws/2005/02/rm"/>
                </sp:SignedParts>
            </wsp:All>
        </wsp:ExactlyOne>
    </wsp:Policy>
    <wsp:Policy wsu:Id="EchoPortBinding_echo_Output_Policy">
        <wsp:ExactlyOne>
            <wsp:All>
```

```
                    <sp:EncryptedParts>
                        <sp:Body/>
                    </sp:EncryptedParts>
                    <sp:SignedParts>
                        <sp:Body/>
                        <sp:Header Name="To"
                                   Namespace="http://www.w3.org/2005/08/addressing"/>
                        <sp:Header Name="From"
                                   Namespace="http://www.w3.org/2005/08/addressing"/>
                        <sp:Header Name="FaultTo"
                                   Namespace="http://www.w3.org/2005/08/addressing"/>
                        <sp:Header Name="ReplyTo"
                                   Namespace="http://www.w3.org/2005/08/addressing"/>
                        <sp:Header Name="MessageID"
                                   Namespace="http://www.w3.org/2005/08/addressing"/>
                        <sp:Header Name="RelatesTo"
                                   Namespace="http://www.w3.org/2005/08/addressing"/>
                        <sp:Header Name="Action"
                                   Namespace="http://www.w3.org/2005/08/addressing"/>
                        <sp:Header Name="AckRequested"
                                   Namespace="http://schemas.xmlsoap.org/ws/2005/02/rm"/>
                        <sp:Header Name="SequenceAcknowledgement"
                                   Namespace="http://schemas.xmlsoap.org/ws/2005/02/rm"/>
                        <sp:Header Name="Sequence"
                                   Namespace="http://schemas.xmlsoap.org/ws/2005/02/rm"/>
                    </sp:SignedParts>
                </wsp:All>
            </wsp:ExactlyOne>
        </wsp:Policy>
    </definitions>
```

The detail is overwhelming. Of interest here is that much of this detail pertains to areas in the SOAP header that need to be encrypted and digitally signed. In any case, a document as complicated as this is best generated using a tool such as the one embedded in the NetBeans IDE rather than written by hand.

The client version, the document *WEB-INF/classes/META-INF/EchoService.xml*, is less complicated:

```
<?xml version="1.0" encoding="UTF-8"?>
<definitions
  xmlns:wsu=
  "http://docs.oasis-open.org/wss/2004/01/oasis-200401-wss-wssecurity-utility-1.0.xsd"
  xmlns:wsp="http://schemas.xmlsoap.org/ws/2004/09/policy"
  xmlns:soap="http://schemas.xmlsoap.org/wsdl/soap/"
  xmlns:tns="http://mcs.ch06/"
  xmlns:xsd="http://www.w3.org/2001/XMLSchema"
  xmlns="http://schemas.xmlsoap.org/wsdl/"
  targetNamespace="http://mcs.ch06/"
  name="EchoService"
  xmlns:sc="http://schemas.sun.com/2006/03/wss/client"
  xmlns:wspp="http://java.sun.com/xml/ns/wsit/policy">
    <wsp:UsingPolicy></wsp:UsingPolicy>
    <types>
        <xsd:schema>
```

```
        <xsd:import
          namespace="http://mcs.ch06/"
          schemaLocation="http://localhost:8080/echoMCS/EchoService?xsd=1">
        </xsd:import>
      </xsd:schema>
    </types>
    <message name="echo">
        <part name="parameters" element="tns:echo"></part>
    </message>
    <message name="echoResponse">
        <part name="parameters" element="tns:echoResponse"></part>
    </message>
    <portType name="Echo">
        <operation name="echo">
            <input message="tns:echo"></input>
            <output message="tns:echoResponse"></output>
        </operation>
    </portType>
    <binding name="EchoPortBinding" type="tns:Echo">
        <wsp:PolicyReference URI="#EchoPortBindingPolicy"/>
        <soap:binding transport="http://schemas.xmlsoap.org/soap/http"
                      style="document"></soap:binding>
        <operation name="echo">
            <soap:operation soapAction=""></soap:operation>
            <input>
                <soap:body use="literal"></soap:body>
            </input>
            <output>
                <soap:body use="literal"></soap:body>
            </output>
        </operation>
    </binding>
    <service name="EchoService">
        <port name="EchoPort" binding="tns:EchoPortBinding">
          <soap:address
            location="http://localhost:8080/echoMCS/EchoService"></soap:address>
        </port>
    </service>
    <wsp:Policy wsu:Id="EchoPortBindingPolicy">
        <wsp:ExactlyOne>
            <wsp:All>
                <sc:CallbackHandlerConfiguration wspp:visibility="private">
                    <sc:CallbackHandler default="wsitUser" name="usernameHandler"/>
                    <sc:CallbackHandler default="changeit" name="passwordHandler"/>
                </sc:CallbackHandlerConfiguration>
                <sc:TrustStore wspp:visibility="private"
                               peeralias="xws-security-server"/>
            </wsp:All>
        </wsp:ExactlyOne>
    </wsp:Policy>
</definitions>
```

The section in bold gives the username, in this case wsitUser, and the password for accessing the server's file-based security realm. The GlassFish command:

```
% asadmin list-file-users
```

can be used to get a list of authorized users. The output should include `wsitUser` for the mutual challenge to succeed.

The servlet-based `Echo` web service and its servlet client `EchoClient` are packaged in separate WAR files and deployed as usual. Here is a dump of each WAR file, starting with the service's *echoMCS.war* (the leftmost column is the file size in bytes):

```
  71 META-INF/MANIFEST.MF
2488 WEB-INF/web.xml
2506 WEB-INF/sun-web.xml
6704 WEB-INF/wsit-ch06.mcs.Echo.xml
 558 WEB-INF/classes/ch06/mcs/Echo.class
```

Although the MCS configuration document has the structure of a WSDL, GlassFish generates the WSDL available to clients in the usual way. The WSDL is essentially the same as *wsit-ch06.mcs.Echo.xml*, but the two do differ in some minor details.

Here is the structure of the client WAR file, *echoClient.war*:

```
  71 META-INF/MANIFEST.MF
 698 WEB-INF/web.xml
 305 WEB-INF/sun-web.xml
 745 WEB-INF/classes/ch06/mcs/client/Echo.class
1583 WEB-INF/classes/ch06/mcs/client/ObjectFactory.class
 646 WEB-INF/classes/ch06/mcs/client/Echo_Type.class
 734 WEB-INF/classes/ch06/mcs/client/EchoResponse.class
 237 WEB-INF/classes/ch06/mcs/client/package-info.class
1484 WEB-INF/classes/ch06/mcs/client/EchoService.class
2042 WEB-INF/classes/ch06/mcs/client/EchoClient.class
 381 WEB-INF/classes/META-INF/wsit-client.xml
2590 WEB-INF/classes/META-INF/EchoService.xml
```

The WAR file includes the *wsimport*-generated classes. The *wsit-client.xml* file does little more than import *EchoService.xml*.

The Dramatic SOAP Envelopes

MCS requires relatively large and complicated SOAP envelopes precisely because the security is now at the SOAP message level rather than at the transport level. The SOAP envelope under MCS needs to include information about encryption algorithms, message digest algorithms, digital signature algorithms, security tokens, time stamps, and so on. All of this information is packaged in a SOAP header block, which is marked with `mustUnderstand` set to `true`; hence, intermediaries and the ultimate receiver must be able to process the header block in some application-appropriate way or else throw a SOAP fault. Here, for dramatic effect, is the request SOAP envelope from a sample `EchoClient` request against the `EchoService`:

```
<?xml version="1.0" ?>
<S:Envelope xmlns:S="http://schemas.xmlsoap.org/soap/envelope/"
            xmlns:wsse="http://docs.oasis-open.org/wss/2004/01/
                oasis-200401-wss-wssecurity-secext-1.0.xsd"
            xmlns:wsu="http://docs.oasis-open.org/wss/2004/01/
```

```
                    oasis-200401-wss-wssecurity-utility-1.0.xsd"
            xmlns:ds="http://www.w3.org/2000/09/xmldsig#"
            xmlns:xenc="http://www.w3.org/2001/04/xmlenc#"
            xmlns:exc14n="http://www.w3.org/2001/10/xml-exc-c14n#">
<S:Header>
  <To xmlns="http://www.w3.org/2005/08/addressing" wsu:Id="_5006">
      http://localhost:8080/echoMCS/EchoService
  </To>
  <Action xmlns="http://www.w3.org/2005/08/addressing"
          wsu:Id="_5005">http://mcs.ch06/Echo/echoRequest
  </Action>
  <ReplyTo xmlns="http://www.w3.org/2005/08/addressing" wsu:Id="_5004">
    <Address>http://www.w3.org/2005/08/addressing/anonymous</Address>
  </ReplyTo>
  <MessageID xmlns="http://www.w3.org/2005/08/addressing"
             wsu:Id="_5003">uuid:b85e5024-85c6-4482-b118-3d00d8ebff17
  </MessageID>
  <wsse:Security S:mustUnderstand="1">
    <wsu:Timestamp xmlns:ns10="http://www.w3.org/2003/05/soap-envelope"
         xmlns:ns11="http://docs.oasis-open.org/ws-sx/ws-secureconversation/200512"
         wsu:Id="_3">
      <wsu:Created>2008-08-14T02:06:32Z</wsu:Created>
      <wsu:Expires>2008-08-14T02:11:32Z</wsu:Expires>
    </wsu:Timestamp>
    <xenc:EncryptedKey xmlns:ns10="http://www.w3.org/2003/05/soap-envelope"
          xmlns:ns11="http://docs.oasis-open.org/ws-sx/ws-secureconversation/200512"
          Id="_5002">
      <xenc:EncryptionMethod
                    Algorithm="http://www.w3.org/2001/04/xmlenc#rsa-oaep-mgf1p"/>
      <ds:KeyInfo xmlns:xsi="http://www.w3.org/2001/XMLSchema-instance"
             xsi:type="keyInfo">
        <wsse:SecurityTokenReference>
          <wsse:KeyIdentifier ValueType="http://docs.oasis-open.org/wss/2004/01/
                oasis-200401-wss-x509-token-profile-1.0#X509SubjectKeyIdentifier"
                EncodingType="http://docs.oasis-open.org/wss/2004/01/
                oasis-200401-wss-soap-message-security-1.0#Base64Binary">
            dVE29ysyFW/iD1la3ddePzM6IWo=
          </wsse:KeyIdentifier>
        </wsse:SecurityTokenReference>
      </ds:KeyInfo>
      <xenc:CipherData>
        <xenc:CipherValue>UIz4WopejXJGmw2ygOMOpIf8hEomI7vI...</xenc:CipherValue>
      </xenc:CipherData>
    </xenc:EncryptedKey>
    <xenc:ReferenceList
        xmlns:ns17="http://docs.oasis-open.org/ws-sx/ws-secureconversation/200512"
        xmlns:ns16="http://www.w3.org/2003/05/soap-envelope"
        xmlns="">
      <xenc:DataReference URI="#_5008"></xenc:DataReference>
      <xenc:DataReference URI="#_5009"></xenc:DataReference>
    </xenc:ReferenceList>
    <xenc:EncryptedData xmlns:ns10="http://www.w3.org/2003/05/soap-envelope"
         xmlns:ns11="http://docs.oasis-open.org/ws-sx/ws-secureconversation/200512"
         Type="http://www.w3.org/2001/04/xmlenc#Element"
         Id="_5009">
```

```
<xenc:EncryptionMethod
    Algorithm="http://www.w3.org/2001/04/xmlenc#aes128-cbc"/>
<ds:KeyInfo xmlns:xsi="http://www.w3.org/2001/XMLSchema-instance"
            xsi:type="keyInfo">
  <wsse:SecurityTokenReference>
    <wsse:Reference ValueType="http://docs.oasis-open.org/wss/
                              oasis-wss-soap-message-security-1.1#EncryptedKey"
                    URI="#_5002"/>
  </wsse:SecurityTokenReference>
</ds:KeyInfo>
<xenc:CipherData>
  <xenc:CipherValue>OGCRGwFKlLfnRYnQd...</xenc:CipherValue>
</xenc:CipherData>
</xenc:EncryptedData>
<ds:Signature xmlns:ns10="http://www.w3.org/2003/05/soap-envelope"
      xmlns:ns11="http://docs.oasis-open.org/ws-sx/ws-secureconversation/200512"
      Id="_1">
  <ds:SignedInfo>
  <ds:CanonicalizationMethod
                      Algorithm="http://www.w3.org/2001/10/xml-exc-c14n#">
    <exc14n:InclusiveNamespaces PrefixList="wsse S"/>
  </ds:CanonicalizationMethod>
  <ds:SignatureMethod Algorithm="http://www.w3.org/2000/09/xmldsig#hmac-sha1"/>
  <ds:Reference URI="#_5003">
    <ds:Transforms>
      <ds:Transform Algorithm="http://www.w3.org/2001/10/xml-exc-c14n#">
        <exc14n:InclusiveNamespaces PrefixList="S"/>
      </ds:Transform>
    </ds:Transforms>
    <ds:DigestMethod Algorithm="http://www.w3.org/2000/09/xmldsig#sha1"/>
    <ds:DigestValue>NI9i+HGoWeYAsu8K1eOcmmSn+SY=</ds:DigestValue>
  </ds:Reference>
  <ds:Reference URI="#_5004">
    <ds:Transforms>
      <ds:Transform Algorithm="http://www.w3.org/2001/10/xml-exc-c14n#">
        <exc14n:InclusiveNamespaces PrefixList="S"/>
      </ds:Transform>
    </ds:Transforms>
    <ds:DigestMethod Algorithm="http://www.w3.org/2000/09/xmldsig#sha1"/>
    ds:DigestValue>5Ab1ebo4/FraGgck/A8iDx1J9+I=</ds:DigestValue>
  </ds:Reference>
  <ds:Reference URI="#_5005">
    <ds:Transforms>
      <ds:Transform Algorithm="http://www.w3.org/2001/10/xml-exc-c14n#">
        <exc14n:InclusiveNamespaces PrefixList="S"/>
      </ds:Transform>
    </ds:Transforms>
    <ds:DigestMethod Algorithm="http://www.w3.org/2000/09/xmldsig#sha1"/>
    <ds:DigestValue>Qso/D/tFg2kzZnbOJ7tOzqRW84M=</ds:DigestValue>
  </ds:Reference>
  <ds:Reference URI="#_5006">
    <ds:Transforms>
      <ds:Transform Algorithm="http://www.w3.org/2001/10/xml-exc-c14n#">
      <exc14n:InclusiveNamespaces PrefixList="S"/>
      </ds:Transform></ds:Transforms>
```

```
                <ds:DigestMethod Algorithm="http://www.w3.org/2000/09/xmldsig#sha1"/>
                <ds:DigestValue>DQsOAHfFqRDBiV4MqOLwbMRLXcc=</ds:DigestValue>
            </ds:Reference>
            <ds:Reference URI="#_5007">
              <ds:Transforms>
                <ds:Transform Algorithm="http://www.w3.org/2001/10/xml-exc-c14n#">
                  <exc14n:InclusiveNamespaces PrefixList="S"/>
                </ds:Transform>
              </ds:Transforms>
              <ds:DigestMethod Algorithm="http://www.w3.org/2000/09/xmldsig#sha1"/>
              <ds:DigestValue>9vpXDjjwI7bLNBAVe5n1jcpHou4=</ds:DigestValue>
            </ds:Reference>
            <ds:Reference URI="#_3">
              <ds:Transforms>
                <ds:Transform Algorithm="http://www.w3.org/2001/10/xml-exc-c14n#">
                  <exc14n:InclusiveNamespaces PrefixList="wsu wsse S"/>
                </ds:Transform>
              </ds:Transforms>
              <ds:DigestMethod Algorithm="http://www.w3.org/2000/09/xmldsig#sha1"/>
              <ds:DigestValue>9NwWEZNcMLbyOfEQlrwbJ6fVGQA=</ds:DigestValue>
            </ds:Reference>
            <ds:Reference URI="#uuid_e2da395f-b9bc-4d52-9cb8-57bafe97ac25">
              <ds:Transforms>
                <ds:Transform Algorithm="http://www.w3.org/2001/10/xml-exc-c14n#">
                  <exc14n:InclusiveNamespaces PrefixList="wsu wsse S"/>
                </ds:Transform>
              </ds:Transforms>
              <ds:DigestMethod Algorithm="http://www.w3.org/2000/09/xmldsig#sha1"/>
              <ds:DigestValue>nRSRynHPET8TPA4DvAR9iB60G1E=</ds:DigestValue>
            </ds:Reference>
          </ds:SignedInfo>
          <ds:SignatureValue>QDgHtRo7NYLsmzKIPDd5RZ/a7hk=</ds:SignatureValue>
          <ds:KeyInfo>
              <wsse:SecurityTokenReference
                          wsu:Id="uuid_5b05ed00-1333-49c3-9f03-0225ea41d3da">
                  <wsse:Reference ValueType="http://docs.oasis-open.org/wss/
                                  oasis-wss-soap-message-security-1.1#EncryptedKey"
                                  URI="#_5002"/>
              </wsse:SecurityTokenReference>
          </ds:KeyInfo>
        </ds:Signature>
      </wsse:Security>
  </S:Header>
  <S:Body wsu:Id="_5007">
    <xenc:EncryptedData xmlns:ns10="http://www.w3.org/2003/05/soap-envelope"
          xmlns:ns11="http://docs.oasis-open.org/ws-sx/ws-secureconversation/200512"
          Type="http://www.w3.org/2001/04/xmlenc#Content" Id="_5008">
      <xenc:EncryptionMethod Algorithm="http://www.w3.org/2001/04/xmlenc#aes128-cbc"/>
      <ds:KeyInfo xmlns:xsi="http://www.w3.org/2001/XMLSchema-instance"
                  xsi:type="keyInfo">
        <wsse:SecurityTokenReference>
          <wsse:Reference
              ValueType=
                  "http://docs.oasis-open.org/wss/oasis-wss-soap-message-security-1.
              1#EncryptedKey" URI="#_5002"/>
```

```
          </wsse:SecurityTokenReference>
        </ds:KeyInfo>
        <xenc:CipherData>
          <xenc:CipherValue>p2TQL4JqgBCWh9Jiv6PWikJObeMuNDvj1wH...</xenc:CipherValue>
        </xenc:CipherData>
      </xenc:EncryptedData>
    </S:Body>
  </S:Envelope>
```

A summary of the under-the-hood steps that produce this SOAP request document should help to clarify the document:

- GlassFish includes a Secure Token Service (*STS*), which is at the center of the mutual authentication between the `EchoService` and its `EchoClient`. The STS receives client requests for security tokens. If a request is honored, the STS sends a *SAML* (Security Assertion Markup Language) token back to the client. SAML is an XML language customized for security assertions. The client can use this token to authenticate itself to a web service. In this example, the `EchoClient` servlet requires a registered username (in this case, `wsitUser`) in order to get a SAML token from the GlassFish STS. This first step establishes trust among the servlet client, the STS, and the web service.

- The servlet client and web service exchange X.509 digital certificates in the process of mutual challenge or authentication. Recall that these certificates include a digital signature, typically from a certificate authority. In test mode, the certificate may be self-signed. The X.509 certificate also includes the public key from the owner's key pair. The role of the digital signature is to vouch for the identity of the public key. This second step establishes a secure conversation between the client and the web service, as each side is now confident about the identity of the other.

- The data exchanged between the client and the web service are packaged, of course, in SOAP envelopes and must be encrypted for confidentiality. In the SOAP request shown earlier, the encrypted data are shown in bold and occur in elements tagged as `xenc:CipherData`

The complexity makes sense given that WSIT cannot rely on the transport level (in particular, HTTPS) for any aspect of the overall security. Instead, WSIT must use the SOAP messages themselves in support of trust, secure conversation, and data confidentiality between the client and the web service. This example drives home the point because the servlet client and the web service do not communicate over HTTPS.

Benefits of JAS Deployment

For Java web services, `Endpoint` is a lightweight publisher; a standalone web container such as Tomcat is a middleweight publisher; and an application server such as GlassFish is a heavyweight publisher. Throughout software, *heavyweight* connotes *complicated*; hence, it may be useful to wind up with a review of what recommends a full application server despite the complications that come with this choice. Here is a short review of the advantages that an application server offers for deploying web services, SOAP-based or REST-style:

- A web service can be implemented as stateless EJB. The EJB container offers services, in particular thread safety and container-managed transactions, that neither `Endpoint` nor a web container offer as is. A production-grade web service is likely to rely on a data store accessible through `@Entity` instances; and `EntityManager` methods such as `find` and `persist` are not inherently thread-safe but become so if the `EntityManager` is instantiated as an EJB field.

- A JAS typically furnishes a web interface to inspect the WSDL and to test any `@WebMethod`. GlassFish provides these and other development tools in its administrative console.

- An application server includes an RDBMS, which is Java DB in the case of Glass-Fish. Other database systems can be plugged in as preferred.

- An enterprise application deployed in an application server can contain an arbitrary mix of components, which may include interactive servlet-based web applications, EJBs, message topics/queues and listeners, `@Entity` classes, and web services. Components packaged in the same EAR file share an *ENC* (Enterprise Naming Context), which makes JNDI lookups straightforward.

- Application servers typically provide extensive administrative support, which includes support for logging and debugging. The GlassFish administrative console is an example.

- GlassFish in particular is the reference implementation of an application server and includes Metro with all of the attendant support for web services interoperability.

- A deployed web service, like any other enterprise component, can be part of a clustered application.

What's Next?

The debate over SOAP-based and REST-style web services is often heated. Each approach has its merits and represents a useful tool in the modern programmer's toolbox. Most SOAP-based services are delivered over HTTP or HTTPS and, as earlier examples show, therefore qualify as a special type of REST-style service. There is no need to pick sides between SOAP and REST. The next and final chapter takes a short look back into the recent history of distributed systems to gain a clearer view of the choices in developing web services.

## CHAPTER 7

Beyond the Flame Wars

The debate within the web services community is often heated and at times intemperate as the advocates of SOAP-based and REST-style services tout the merits of one approach and rail against the demerits of the other. Whatever the dramatic appeal of a SOAP versus REST conflict, the two approaches can coexist peacefully and productively. There is no hard choice here—no *either one or the other but not both*. Each approach has its appeal, and either is better than legacy approaches to distributed software systems. A quick look at the history of distributed systems is one way to gain perspective on the matter.

A Very Short History of Web Services

Web services evolved from the *RPC* (Remote Procedure Call) mechanism in *DCE* (Distributed Computing Environment), a framework for software development that emerged in the early 1990s. DCE included a distributed file system (DCE/DFS) and a Kerberos-based authentication system. Although DCE has its origins in the Unix world, Microsoft quickly did its own implementation known as *MSRPC*, which in turn served as the infrastructure for interprocess communication in Windows. Microsoft's *COM/ OLE* (Common Object Model/Object Linking and Embedding) technologies and services were built on a DCE/RPC foundation. The first-generation frameworks for distributed-object systems, *CORBA* (Common Object Request Broker Architecture) and Microsoft's *DCOM* (Distributed COM), are anchored in the DCE/RPC procedural framework. Java RMI also derives from DCE/RPC, and the method calls in the original EJB types, Session and Entity, are Java RMI calls. Java EE and Microsoft's .NET, the second-generation frameworks for distributed-object systems, likewise trace their ancestry back to DCE/RPC. Various popular system utilities (for instance, the Samba file and print service for Windows clients) use DCE/RPC.

DCE/RPC has the familiar client/server architecture in which a client invokes a procedure that executes on the server. Arguments can be passed from the client to the server and return values can be passed from the server to the client. The framework is language-neutral in principle, although strongly biased toward C/C++ in practice.

DCE/RPC includes utilities for generating client and server artifacts (stubs and skeletons, respectively) and software libraries that hide the transport details. Of interest now is the *IDL* (Interface Definition Language) document that acts as the service contract and is an input to utilities that generate artifacts in support of the DCE/RPC calls.

The Service Contract in DCE/RPC

Here is a simple IDL file:

```
/* echo.idl */
[uuid(2d6ead46-05e3-11ca-7dd1-426909beabcd), version(1.0)]
interface echo
{
    const long int ECHO_SIZE = 512;
    void echo(
        [in]            handle_t h,
        [in,  string]   idl_char from_client[ ],
        [out, string]   idl_char from_server[ECHO_SIZE]
    );
}
```

The interface, identified with a machine-generated UUID, declares a single function of three arguments, two of which are in parameters (that is, inputs to the remote procedure) and one of which is an out parameter (that is, an output from the remote procedure). The first argument, of defined type handle_t, is required and points to an RPC binding structure. The function echo could but does not return a value because the echoed string is returned instead as an out parameter. The IDL, though obviously not in XML format, is functionally akin to a WSDL.

The IDL file is passed through an IDL compiler (for instance, the Windows *midl* utility) to generate the appropriate client-side or server-side artifacts. Here is a C client that invokes the echo procedure remotely:

```
#include <stdio.h>   /* standard C header file */
#include <dce/rpc.h> /* DCE/RPC header file */

int main(int argc, char* argv[ ]) {
    /* DCE/RPC data types */
    rpc_ns_handle_t import_context;
    handle_t        binding_h;
    error_status_t  status;
    idl_char        reply[ECHO_SIZE + 1];

    char* msg = "Hello, world!";

    /* Set up RPC. */
    rpc_ns_binding_import_begin(rpc_c_ns_syntax_default,
                                (unsigned_char_p_t) argv[1], /* server id */
                                echo_v1_0_c_ifspec,
                                0,
                                &import_context,
                                &status);
```

```
        check_status_maybe_die(status, "import_begin"); /* code not shown */

        rpc_ns_binding_import_next(import_context,
                                   &binding_h,
                                   &status);
        check_status_maybe_die(status, "import_next");

        /* Make the remote call.  */
        echo(binding_h, (idl_char*) msg, reply);
        printf("%s echoed from server: %s\n", msg, reply);

        return 0;
    }
```

The setup code at the start is a bit daunting, but the call itself is straightforward.

The IDL file plays a central role in Microsoft's ActiveX technology. An ActiveX control is basically a *DLL* (Dynamic Link Library) with an embedded *typelib*, which is a compiled IDL file. So an ActiveX control is a DLL that carries along its own RPC interface, which means that the control can be embedded dynamically in a host application (for instance, a browser) that can consume the typelib to determine how to invoke the calls implemented in the DLL. This is a preview of how the client of a SOAP-based web service will consume a WSDL.

DCE/RPC is an important step toward language-agnostic distributed systems but has some obvious, if understandable, drawbacks. For one, DCE/RPC uses a proprietary transport, which is understandable in a pioneering technology that predates HTTP by almost a decade. My point is not to evaluate DCE/RPC but rather to highlight similarities between it and its descendant, SOAP-based web services. To that end, XML-RPC is the next SOAP ancestor to consider.

XML-RPC

Dave Winer of UserLand Software developed XML-RPC in the late 1990s. XML-RPC is a very lightweight RPC system with support for elementary data types (basically, the built-in C types together with a `boolean` and a `datetime` type) and a few simple commands. The original specification is about seven pages in length. The two key features are the use of XML marshaling/unmarshaling to achieve language neutrality and reliance on HTTP (and, later, SMTP) for transport. The O'Reilly open-wire Meerkat service is an XML-RPC application.

As an RPC technology, XML-RPC follows the request/response pattern. Here is the XML request from an invocation of the Fibonacci function with an argument of 11, which is passed as a four-byte integer:

```
<?xml version="1.0"?>
<methodCall><methodName>ch07.fib</methodName>
  <params><param><value><i4>11</i4></value></param></params>
</methodCall>
```

XML-RPC is deliberately low fuss and very lightweight. For an XML-RPC fan, SOAP is XML-RPC after a prolonged eating binge.

Standardized SOAP

Perhaps the most compelling way to contrast the simplicity of XML-RPC with the complexity of SOAP is to list the categories for SOAP-related standards initiatives and the number of initiatives in each category. (An amusing poster of these standards is available at *http://www.innoq.com/resources/ws-standards-poster*.) Table 7-1 shows the list.

Table 7-1. Standards for SOAP-based web services

Category	Number of standard initiatives
Interoperability	10
Security	10
Metadata	9
Messaging	9
Business Process	7
Resource	7
Translation	7
XML	7
Management	4
SOAP	3
Presentation	1

The total of 74 distinct but related initiatives helps explain the charge that SOAP-based web services are over-engineered. Some of the standards (for instance, those in the Business Process category) represent an effort to make web services more useful and practical. Others (for instance, those in the categories of Interoperability, Security, Metadata, and SOAP itself) underscore the goal of language, platform, and transport neutrality for SOAP-based services. Recall the examples in Chapter 6 of mutual challenge security (MCS) using digital certificates. If the MCS is done under HTTPS, the setup is minimal and the exchanged SOAP envelopes carry no security information whatever. If the MCS is done under WSIT, the setup is tricky and the exchanged SOAP envelopes have relatively large, complicated header blocks that encapsulate security credentials and supporting materials. However, WSIT complexity can be avoided, in this case, by letting HTTPS transport carry the MCS load. Despite the complexities, even SOAP-based services are an evolutionary advance.

SOAP-Based Web Services Versus Distributed Objects

Java RMI (including the Session and Entity EJB constructs built on it) and .NET Remoting are examples of second-generation distributed object systems. Consider what a Java RMI client requires to invoke a method declared in a service interface such as this:

```
package ch07.doa;  // distributed object architecture
import java.util.List;
public interface BenefitsService extends java.rmi.Remote {
    public List<EmpBenefits> getBenefits(Emp emp) throws RemoteException;
}
```

The client needs a Java RMI *stub*, an instance of a class that implements the interface `BenefitsService`. The stub is downloaded automatically to the client through serialization over the socket that connects the client and the service. The stub class requires the programmer-defined `Emp` and `EmpBenefits` types, which in turn may nest other programmer-defined types, e.g., `Department`, `BusinessCertification`, and `ClientAccount`:

```
public class Emp {
    private Department                    department;
    private List<BusinessCertification>   certifications;
    private List<ClientAccount>           accounts;
    ...
}
```

The standard Java types such as `List` already are available on the client side, as the client is, by assumption, a Java application. The challenge involves the programmer-defined types such as `Emp` and `EmpBenefits` that are needed to support the client-side invocation of a remotely executed method, as this code segment illustrates:

```
List<EmpBenefit> fred_benefits = remote_object.getBenefits(fred);
```

Under Java RMI, programmer-defined types required on the client side are loaded remotely and automatically. Yet the remote loading of Java classes is inherently complicated and, of course, the format of Java's *.class* files is proprietary. The point is that lots of stuff needs to come down from the server to the client before the client can invoke a method remotely. The arguments passed to the Java RMI service (in this example, the `Emp` instance to which `fred` refers) are serialized on the client side and deserialized on the service side. The returned value, in this case a list of `EmpBenefits`, is serialized by the service and deserialized by the client. The client and service communicate through binary rather than text streams, and the structure of the binary streams is specific to Java.

SOAP-based web services simplify matters. For one thing, the client and service now exchange XML documents: *text* in the form of SOAP envelopes. (In special cases, raw bytes can be exchanged instead.) The exchanged text can be inspected, validated, transformed, persisted, and otherwise processed using readily available and often free tools. Each side, client and service, simply needs a local software library that binds language-specific types such as the Java `String` to XML Schema (or comparable) language-neutral types, in this case `xsd:string`. Given these bindings, relatively simple library modules can serialize and deserialize. Processing on the client side, as on the service side, requires only *locally* available libraries and utilities. The complexities can be isolated at the endpoints and need not seep into the exchanged SOAP messages.

As noted repeatedly, the client and the service need not be coded in the same language or even in the same style of language. Clients and services can be implemented in object-oriented, procedural, functional, and other language styles. The languages on either end may be statically typed (for instance, Java and C#) or dynamically typed (for example, Perl and Ruby), although it is easier to generate a WSDL from a service coded in a statically typed language. Further, SOAP insulates the client and the service from platform-specific details. Life gets simpler and proportionately better with the move from distributed objects to SOAP-based web services.

SOAP and REST in Harmony

From a programmer's standpoint, there are two profoundly different ways to do SOAP-based web services. The first way, which dominates in practice, is to assume HTTP/HTTPS transport and to let the transport carry as much of the load as possible. In this approach, the transport handles requirements such as reliable messaging, peer authentication, and confidentiality. The second way, which accounts for much of the complexity in the WS-* initiatives, requires the programmer to deal with the many APIs that address transport-neutral services.

The first way is certainly my preferred way. If HTTP/HTTPS is assumed as transport, then SOAP-based web services in the predominant request/response pattern are just a special case of RESTful services: a client request is a POSTed SOAP envelope, and a service response is likewise a SOAP envelope. The very good news is that the SOAP envelopes are mostly transparent to the programmer, who does not have to generate the XML on the sending side or to parse the XML on the receiving side. The client examples in Chapter 4, on RESTful services, illustrate the point. In every client, the bulk of the code is devoted to XML processing. *Transparent XML* sounds like an empty slogan only to someone who has not had to deal directly with large and complicated XML documents.

From the programmer's viewpoint, the SOAP-based approach also has the enormous appeal of a service contract, a WSDL, that can be used to generate client-side artifacts that standardize the task of writing the client. Indeed, the major drawback of RESTful services in general is the lack of a uniform service contract that facilitates coding the client. An XML Schema is a grammar, not a service contract. Of course, the grammar can guide the programmer in parsing a response XML document to extract the items of interest; but this is little consolation to the programmer if every RESTful service has its own idiosyncratic response document, not to mention its own distinctive invocation syntax. The next big step for the RESTful community is to agree on a format, such as the WADL document, to act as a service contract. Imagine the boost that the RESTful style would enjoy from a lightweight service contract that could be used to generate client-side artifacts.

SOAP-based and REST-style services present different challenges but likewise offer different opportunities. The SOAP-based approach mostly hides the XML, whereas the REST-style approach typically requires that the XML be front and center. The SOAP-based approach does not take direct advantage of what HTTP offers, although the SOAP in SOAP-based services is almost always delivered over HTTP/HTTPS. The RESTful approach obviously exploits what HTTP/HTTPS offers but, so far, does little to ease the task of processing the XML payloads.

Let's hope that a unified approach to RESTful services will be forthcoming. Until then my two cents still will be spent on RESTful services—and on SOAP-based ones delivered over HTTP/HTTPS.

Index

We'd like to hear your suggestions for improving our indexes. Send email to *index@oreilly.com*.

portType section (WSDL), 37
POST HTTP verb, 122, 136
pre-master secret, 199
primitive data types, 67
private keys, 197
programmatic security, 220
public keys, 197
    certificates, 198
    cryptography algorithms, 201
publish/subscribe messaging, 262
publishers, 239
publishing web services, 6
PUT HTTP verb, 122, 136

Q

QName constructor, 31, 34
qualified names (XML), 14
query strings, 125
queue-based messaging, 262

R

Rails, 186
raw XML, 125, 177
RC4_128, 201
RDMS (Relational Database Management
    System), 242
realms, 219
receivers, 239
reference implementation (RI), 5, 239
Relational Database Management System
    (RDMS), 242
Remote Method Invocation (RMI), 240, 287
Remote Procedure Call (see rpc style)
REpresentational State Transfer (see REST)
requesters
    Java, 13
    Perl/Ruby, 10–11
@Resource, 263
REST (REpresentational State Transfer), 2,
    121–191
    Amazon Associates Web Service and, 46
    HttpServlet class, 159–167
    language transparency and, 132–136
RESTful web services, 121–191
    (see also REST)
Restlet framework, 186–191
RI (reference implementation), 239
Rivest, Shamir, Adleman (RSA), 201

RMI (Remote Method Invocation), 240, 287
Ron's Code, 201
root elements (WSDL), 37
rpc attribute (SOAP), 38
rpc style, 41, 42, 283
    document styles and, 55
    wsgen utility and, 60
RSA (Rivest, Shamir, Adleman), 201
Ruby, 2, 14
    requesters of web services, 10–11

S

SAML (Security Assertion Markup Language),
    280
SAX (Simple API for XML), 136
Schema (XML), 37
    Java types and, 67
    SOAP messages, validating, 42
secret access keys, 46
secret keys, 196
Secure Hash Algorithm-1 (SHA-1), 84
Secure Token Service (STS), 280
security, 193–238
    providers, 241
    WS-Security and, 227–238
Security Assertion Markup Language (SAML),
    280
SEI (Service Endpoint Interface), 4
    client-support code and, 32
serialization, 62
service contract, 17
Service Endpoint Interface (see SEI)
service endpoints, 9
service implementation (UDDI), 79
service implementation bean (see SIB)
service interface (UDDI), 79
service section (WSDL), 38
service-side SOAP handlers, 97
Service.create method, 31
Service.Mode.PAYLOAD, 159
servlet-based implementations, 261
set method, 28, 64
SH (SOAP handlers), 89
SHA-1 (Secure Hash Algorithm-1), 84
SIB (service implementation bean), 31
    MessageContext and, 104
Simple Mail Transfer Protocol (SMTP), 17, 37
Simple Object Access Protocol (see SOAP)
single (secret) key, 196

Y

Yahoo! news service, 167–170

About the Author

Martin Kalin has a Ph.D. from Northwestern University and is a professor in the College of Computing and Digital Media at DePaul University. He has cowritten a series of books on C and C++ and has written a book on Java for programmers. He enjoys commercial programming and has codeveloped large distributed systems in process scheduling and product configuration.

Colophon

The animal on the cover of *Java Web Services: Up and Running* is a great cormorant (*Phalacrocorax carbo*). This bird is a common member of the Phalacrocoracidae family, which consists of about 40 species of cormorants and shags—large seabirds with hooked bills, colored throats, and stiff tail feathers. The name "cormorant" is derived from the Latin *corvus marinus*, or sea raven, because of its black plumage.

An adult cormorant is about 30 inches long and has a wingspan of 60 inches. It has a long neck, a yellow throat, and a white chin patch. The cormorant has a distinctive way of spreading its wings as it perches, which many naturalists believe is a way of drying its feathers, although this explanation is disputed. The cormorant lives mainly near Atlantic waters, on western European and eastern North American coasts, and particularly on the Canadian maritime provinces. It breeds on cliffs or in trees, building nests out of twigs and seaweed.

The cormorant is an excellent fisher, able to dive to great depths. In China and Japan, among other places, fishermen have trained this bird using a centuries-old method, in which they tie cords around the throats of the birds—to prevent them from swallowing—and send them out from boats. The cormorants then catch fish in their mouths, return to the boats, and the fishermen retrieve their catch. Although once a successful industry, today cormorant fishing is primarily used for tourism.

The cover image is from Cassell's *Popular Natural History, Vol. III: Birds*. The cover font is Adobe ITC Garamond. The text font is Linotype Birka; the heading font is Adobe Myriad Condensed; and the code font is LucasFont's TheSansMonoCondensed.

Related Titles from O'Reilly

Java

Ajax on Java

Ant: The Definitive Guide, *2nd Edition*

Better, Faster, Lighter Java

Beyond Java

Eclipse

Eclipse Cookbook

Eclipse IDE Pocket Guide

Enterprise JavaBeans 3.0, *5th Edition*

Hardcore Java

Head First Design Patterns

Head First Design Patterns Poster

Head First Java, *2nd Edition*

Head First Servlets & JSP

Head First EJB

Hibernate: A Developer's Notebook

J2EE Design Patterns

Java 5.0 Tiger: A Developer's Notebook

Java & XML Data Binding

Java & XML, *3rd Edition*

Java Cookbook, *2nd Edition*

Java Data Objects

Java Database Best Practices

Java Enterprise Best Practices

Java Enterprise in a Nutshell, *3rd Edition*

Java Examples in a Nutshell, *3rd Edition*

Java Extreme Programming Cookbook

Java Generics and Collections

Java in a Nutshell, *5th Edition*

Java I/O, *2nd Edition*

Java Management Extensions

Java Message Service

Java Network Programming, *3rd Edition*

Java NIO

Java Performance Tuning, *2nd Edition*

Java RMI

Java Security, *2nd Edition*

JavaServer Faces

JavaServer Pages, *3rd Edition*

Java Servlet & JSP Cookbook

Java Servlet Programming, *2nd Edition*

Java Swing, *2nd Edition*

Java Web Services in a Nutshell

JBoss: A Developer's Notebook

JBoss at Work: A Practical Guide

Learning Java, *3rd Edition*

Mac OS X for Java Geeks

Maven: A Developer's Notebook

Programming Jakarta Struts, *2nd Edition*

QuickTime for Java: A Developer's Notebook

Spring: A Developer's Notebook

Swing Hacks

Tomcat: The Definitive Guide, *2nd Edition*

WebLogic: The Definitive Guide